Controlled Coupling of a Single Nanoparticle to a High-Q Microsphere Resonator

Dissertation

zur Erlangung des akademischen Grades
doctor rerum naturalium
(Dr. rer. nat.)
im Fach Physik
eingereicht an der

Mathematisch-Naturwissenschaftlichen Fakultät I
der Humboldt-Universität zu Berlin

von

Stephan Götzinger
geboren am 29.09.1973 in Kaiserslautern

Präsident der Humboldt-Universität zu Berlin
Prof. Dr. Jürgen Mlynek

Dekan der Mathematisch-Naturwissenschaftlichen Fakultät I
Prof. Dr. Michael Linscheid

Diese Arbeit wurde von der Mathematisch-Naturwissenschaftlichen Fakultät I der Humboldt-Universität zu Berlin 2004 als Dissertation angenommen.

Bibliografische Information Der Deutschen Bibliothek

Die Deutsche Bibliothek verzeichnet diese Publikation in der Deutschen Nationalbibliografie; detaillierte bibliografische Daten sind im Internet über http://dnb.ddb.de abrufbar.

ISBN 3-8325-0513-X

Logos Verlag Berlin
Comeniushof, Gubener Str. 47,
10243 Berlin
Tel.: +49 030 42 85 10 90
Fax: +49 030 42 85 10 92
INTERNET: http://www.logos-verlag.de

Zusammenfassung

Im Rahmen dieser Arbeit konnte erstmals die gezielte Kopplung eines einzelnen Nanoteilchens an die Moden eines Mikrokugelresonators hoher Güte realisiert werden.
Dazu wurde zunächst eine neuartige Spektroskopie-Einheit konstruiert. Die gemessenen Güten der selbsthergestellten Mikrokügelchen ($40 - 300\,\mu m$) überstiegen 10^9, was bei Kugeldurchmessern kleiner $140\,\mu m$ einer Finesse von mehr als 10^6 entspricht. Um die sogenannten "Whispering-gallery" Moden des Mikrokugelresonators eindeutig indentifizieren zu können, wurde ein optisches Nahfeldmikroskop entwickelt und aufgebaut. Mit diesem konnte die Intensitätsverteilung der Moden auf der Kugeloberfäche mit einer Auflösung unterhalb des Beugungslimits abgebildet und so die Orientierung der Mode im Raum exakt ermittelt werden. Dadurch wurde es zum ersten Mal möglich, eine Optimierung der schwierigen Kopplung an die Fundamentalmode (das ist die Mode mit dem kleinsten Modenvolumen) auf gezielte Art und Weise vorzunehmen. Die Moden und deren Kopplung waren nun durch den Einsatz des Nahfeldmikroskopes vollständig beherrscht.
In einem nächsten Schritt wurde der Einfluß der Nahfeldsonde auf die Güte des Resonators untersucht. Dieser ist von entscheidender Bedeutung beispielsweise für die Realisierung eines Lasers mit nur einem Quantenemitter als aktivem Material. Bei einem solchen Laser wird der Nanoemitter von außen an das Mikrokügelchen angenähert. Ausschlaggebend ist dabei, ob der Emitter im Maximum der Fundamentalmode positioniert werden kann und gleichzeitig Güten von etwa 10^8 gehalten werden können. Dazu wurden systematische Messungen mit Nahfeldsonden verschiedener Größe durchgeführt. Es zeigte sich, daß Spitzengrößen von etwa $100\,nm$ keinen messbaren Einfluß auf die Güte des Resonators haben, selbst wenn diese größer als 10^8 ist. Solche Nahfeldsonden sind also ideal geeignet, einzelne Quantenemitter mit Nanometerpräzision im Feld einer Mode hoher Güte zu positionieren.
Schließlich wurde die Kopplung einzelner Nanoteilchen ($100 - 500\,nm$ große farbstoffdotierte Kolloide) realisiert. Dabei wurden zwei Wege parallel beschritten: Zum einen wurde ein eigens dafür entwickeltes "beam scanning" Konfokalmikroskop mit Einzelmolekülsensitivität verwendet, um gezielt einzelne Teilchen auf der Oberfläche des Mikrokügelchens anzuregen. Zum anderen konnte ein Nanoteilchen, welches am Ende einer Nahfeldsonde befestigt war, mit Hilfe des Nahfeldmikroskopes kontrolliert an die Moden des Resonators gekoppelt werden. Unter Verwendung des Konfokalmikroskopes konnte sogar zum ersten Mal die Kopplung eines einzelnen Quantenemitters (CdSe/ZnS Nanokristall) nachgewiesen werden.
Am Ende der Arbeit wurden erste Experimente durchgeführt, welche sich mit der Kopplung von zwei Nanoteilchen über die Moden des Resonators beschäftigen. Vielversprechende Resultate konnten hier bereits erzielt werden. Die ersten experimentellen Ergebnisse weisen auf die Kopplung zweier $200\,nm$ großer Kolloide durch Photonenaustausch über gemeinsame Resonatormoden hin.

Abstract

In this work, the controlled coupling of a single nanoparticle to a high-Q microsphere resonator was realized for the first time.
A novel optical experiment including a grating stabilized diode laser was established to conduct spectroscopy on microspheres. The measured Q-factors of the fabricated microspheres (diameter $40 - 300\,\mu m$) exceeded 10^9, which corresponds for sphere sizes smaller than $140\,\mu m$ to a Finesse larger than 10^6. For unequivocal identification of the so-called high-Q whispering-gallery modes of the microsphere, an optical near-field microscope was developed, by which the intensity distribution of the modes on the sphere's surface could be imaged beyond the diffraction limit. Comparison of the intensity distribution with the microsphere's topography, which is given by the shear-force feedback loop of the near-field probe, makes it possible to determine the exact orientation of the mode in space. In this way it was possible for the first time to optimize the difficult coupling to the fundamental mode (which is the mode with the smallest mode volume) in a well-defined manner. Therefore, the modes and their coupling were completely controlled by the use of the near-field microscope.
In a next step, the influence of a near-field probe on the Q-factor of a microsphere resonator was investigated. This is of great importance for the realization of a laser, for example, with only a single quantum emitter as active material. In such a laser the emitter is attached to the near-field probe and approaches the microsphere from outside. It is essential that the emitter can be placed in the maximum of the field of the fundamental mode and that Q-factors of about 10^8 can be maintained at the same time. Systematic studies with near-field probes of different sizes were performed. Experimental results showed that probes with a tip size of about $100\,nm$ did not show any measurable influence on the Q-factor, even when the Q was larger than 10^8. Such near-field probes are therefore perfect nanohandles which can be used to place a single quantum emitter (e.g. a molecule or a semiconductor quantum dot) with nanometer precision in the field of a high-Q mode.
Finally, the coupling of single nanoparticles was realized. Dye-doped beads with a size of $100 - 500\,nm$ or CdSe/ZnS nanocrystals were used. Two strategies were employed in parallel. A specially designed beam scanning confocal microscope was used to address single nanoparticles independently on the surface of a microsphere. The coupling of a single nanocrystal as a true quantum emitter was demonstrated for the first time. In the second approach a nanoparticle was attached to the end of a near-field probe. Since the particle could now be moved with nanometer precision on the sphere's surface, it could be coupled in a controlled way to the modes of the resonator.
At the end of this work, experiments concerning the coupling of two nanoparticles via the resonator modes were started, and promising results were obtained. There is strong evidence of the coupling of two $200\,nm$ dye-doped beads via photon exchange of shared high-Q modes.

Contents

Chapter 1

Introduction

A well-known acoustic phenomenon can be demonstrated impressively in St. Paul's Cathedral in London. At the base of the Cathedral's dome is a walkway which forms a circular gallery 40 m above the main floor, with a diameter of about 33 m (see figure 1.1 a)). When someone whispers along the gallery wall they can be heard by anyone who is close to the wall at any point around the walkway. However, if the speaker whispers directly to the center of the gallery, nothing can be heard. In the 19th century this audible anomaly was brought to scientific attention [Wal78]. Lord Rayleigh was the first who recognized that sound waves are guided by the walls of the gallery [Ray78]. In particular, he realized that a whisper along the walls of the gallery can excite the acoustic eigenmodes of the circular dome [Ray10]. These eigenmodes carry sound with a clearly reduced damping. The acoustic phenomenon of the "whispering-gallery" has been associated with other effects occurring in physics, e.g. in superconductivity [WUK$^+$00] and optics [GKL61].

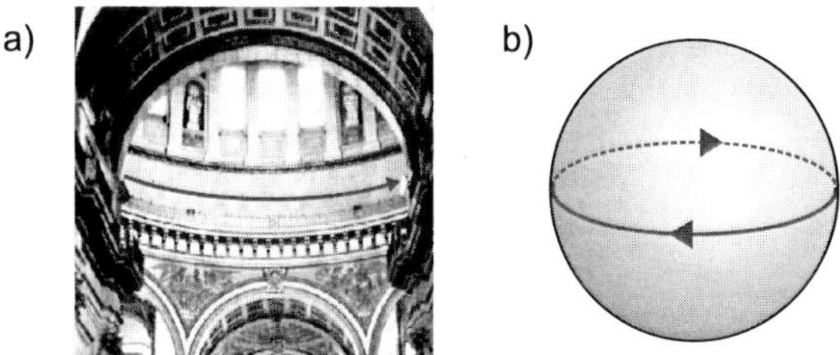

Figure 1.1: *Analogy between acoustical and optical whispering-gallery modes: a) the whispering-gallery in the dome of St. Paul's Cathedral [WU02]. b) Sketch of a microsphere: light is guided inside by total internal reflection.*

In 1989, a Russian group reported on an optical microcavity based on the "whispering-gallery" effect [BGI89]. With a small silica sphere (size about 100 μm), it was possible to trap visible light in a small volume (see figure 1.1 b)), better than in any other confined geometry[1]. Already in this first publication, the authors mentioned possible applications of these microspheres: micro-optic filtering and narrow line integrated microlasers. In addition, the possibility of using these microspheres as the basic element in an optical computer was also discussed, because of their optical switching capabilities.

The property of storing light (expressed in the Q-factor of a cavity) associated with a small mode volume makes these spheres extremely attractive for a branch of physics called cavity **Q**uantum **E**lectro**D**ynamics (QED) [VFG+98]. A fundamental observation of cavity QED was initiated by Purcell in 1946 [Pur46]. The lifetime of an excited atom is not an intrinsic property of the atom itself, but depends also on its environment. In other words, the modification of the mode structure of the electromagnetic vacuum field by an optical cavity influences the spontaneous decay of an excited state (Purcell effect). Even more pronounced are cavity QED effects in the so-called strong coupling regime. If the atomic transition is in resonance with a cavity eigenmode, then the atom and the field-mode exchange one energy quantum periodically with the so-called Rabi-frequency. The irreversible spontaneous decay changes its character completely into a coherent oscillatory behavior. This happens when the atom cavity coupling strength is faster than any underlying dissipative rate [Vah03], e.g. the spontaneous decay time of the atom or the cavity decay time. Thus both the Purcell effect and strong coupling depend on the Q-factor and the mode volume, as well as the position of the atom (or any other quantum emitter) relative to the mode. The latter determines the coupling strength and therefore its control is of crucial importance for the observation of these phenomena.

Cavity QED effects were first observed in atomic physics experiments [Kle81, GRGH83, RTB+91]. Here a moderate control of the position of the atoms is obtained by a tightly focussed atomic beam or in more recent experiments [GKH+01, GRL+03] by the use of ion traps (see figure 1.2). However, when the extremely high-Q of a microsphere resonator is exploited, problems arise. The emitter needs to be located a few nanometers above the sphere's surface to achieve sufficient coupling to the whispering-gallery modes. Such a realization, with atoms or ions, is confronted by various experimental challenges, which have not been addressed, e.g the adsorption of atoms on the sphere's surface.

On the other hand, choosing solid state emitters, such as a single quantum dot [BBA+92] or a single molecule [MK89] makes it possible to position the emitter with sub-wavelength precision within the cavity using modern micro- and nanostructuring techniques. Recently, experiments where quantum dots were implanted in semiconductor resonators were carried out [PLS+99, MKB+00, SPY01]. However, once the quantum dots have been implanted, their positions are fixed and it is impossible to change or optimize the coupling to certain electromagnetic modes. In order to study the interaction between a single quantum emitter and a cavity mode in detail, many

[1] If the whispering-gallery of St. Paul's Cathedral had the same storage capability for sound, e.g for the 440 Hz standard pitch, one could hear a tone produced once for about one month.

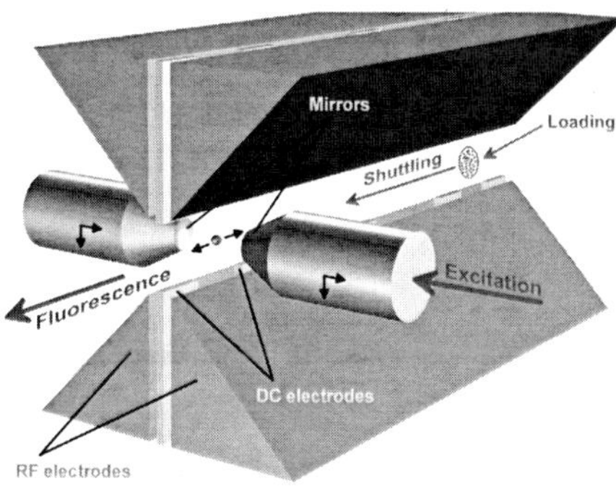

Figure 1.2: *Example of a cavity QED experiment in atomic physics: a single ion is trapped in a linear ion trap and transported into a high Finesse Fabry-Perot cavity [GKH+01].*

different samples have to be grown in order to vary a certain parameter, for instance the absolute position of the quantum dot in the cavity. Even then, the finite reproducibility of the structure fabrication makes it difficult to distill the influence of a certain parameter from the experimental data.
The ideal situation, where the emitter can be placed with nanometer accuracy at any desired position, could be achieved by combining solid state emitters with scanning probe techniques. In figure 1.3 an experimental setup using microspheres is proposed, in which one would have full control over the coupling conditions. Such a setup would

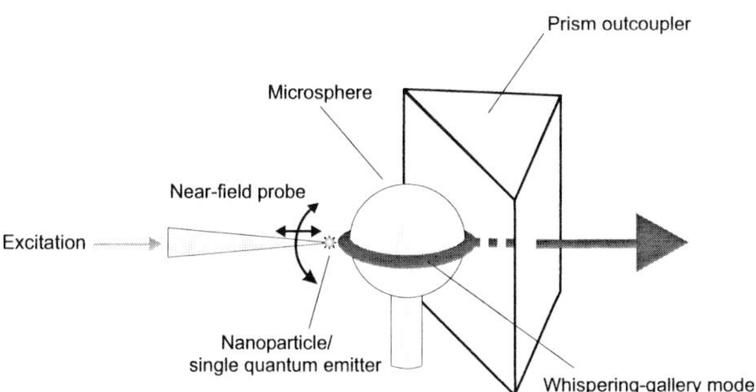

Figure 1.3: *Experimental setup for realization of a controlled coupling of a single nanoparticle to a high-Q microsphere resonator.*

allow the realization of a wide variety of experiments, where a single or a few nanoscopic emitters such as molecules or quantum dots would interact with single high-Q modes

of the electromagnetic field. Even the realization of novel light sources such as a single quantum emitter laser might become feasible [PY99, BY99].
In this work, methods have been developed that aim at the optimal realization of the coupling between a single nanoparticle and a single mode of a microsphere resonator with a low mode volume and a very high Q-factor.
The main part of this work consists of eight chapters. A short description of each chapter is given below.

Chapter 2. This chapter introduces the concepts of high-Q whispering-gallery modes present in silica microspheres. Almost all equations, symbols and concepts needed later in this work are introduced here.

Chapter 3. For this work several experimental tools must be developed. An entire chapter is dedicated to their technical peculiarities. The chapter describes the building of a grating stabilized diode laser for whispering-gallery mode spectroscopy, the development of a beam scanning confocal microscope and also the construction of a scanning near-field optical microscope.

Chapter 4. The fabrication of high-Q silica microspheres and their spectroscopy are described in this chapter. Furthermore, the use of a near-field microscope is demonstrated for unequivocal identification of whispering-gallery modes. A procedure is developed to optimize the coupling to the fundamental mode, that is the mode with the smallest mode volume.

Chapter 5. Here, the influence of near-field probes with different sizes on the Q-factor of a microsphere is studied. This is of special importance if one tries to couple some active material to the sphere by placing the emitter into the evanescent field outside of the sphere. Any influence of the handle where the emitter is attached has to be minimized.

Chapter 6. This chapter deals with the most important results of this work. A single nanocrystal is excited with the beam scanning confocal microscope and its coupling to the microsphere is demonstrated. In another experiment, a single nanoparticle is attached to the end of a near-field probe. The use of scanning probe techniques gives ultimate control over the coupling conditions.

Chapter 7. Initial results concerning the coupling of two nanoparticles via whispering-gallery modes are shown in this chapter. Also, an elegant scheme using an upconversion process for these kind of experiments is described.

Chapter 8. Finally, additional experiments which can be realized using the knowledge developed during this work are discussed. Especially the implementation of the setup in a cryogenic environment should allow one to perform exciting experiments in the field of cavity QED, as well as the realization of an unusual laser with only a few or even a single quantum emitter as active material.

Chapter 2

Microspheres as optical resonators

This chapter introduces the main concepts of high-Q whispering-gallery modes, sometimes also called morphology dependent resonances, present in microsphere resonators. Whispering-gallery modes belong to the wider family of Mie resonances, theoretically already described in 1908 by Gustav Mie in his pioneering work on the scattering of colloidal metal solutions [Mie08]. Since then scattering of light by small spheres has been studied extensively both experimentally and theoretically [Str41, BC88, BH98], up to a level of mathematical sophistication where microspheres are treated as open systems with quasinormal modes [CC96]. However, using full Mie theory is algebraically and even computationally cumbersome; it often tends to obscure simple ideas behind lengthy and unhandy formulas. The scope of this chapter is to provide the reader with a fundamental knowledge of whispering-gallery modes, based on intuitive pictures rather than on a rigorous mathematical treatment found in the references mentioned above. The quality factor will be introduced first, as the most important property of microspheres. Then, in section 2.2 the analytical field distribution of a mode inside a sphere is derived, following Stratton's book [Str41]. Usually there are different approaches used to explain the physical ideas behind the whispering-gallery modes and their interpretation, e.g. geometrical optics, precession of gaussian beams inside the sphere, series expansions and other approximations, all with the aim of giving a physical interpretation of these modes. In the remaining subsections, almost all of these concepts are used, not only to give the reader a first insight into the general culture of the field, but also because the different approaches provide a deeper understanding.

2.1 Quality factor of a cavity

The most prominent characteristic of a silica microsphere resonator is the extremely high quality factor (Q-factor). The Q-factor of a cavity is a measure of the losses, defined as 2π times the stored energy in the cavity divided by the energy losses per cycle [Jac98].

$$Q = \omega_0 \frac{\text{Stored energy}}{\text{Power loss}} \tag{2.1}$$

Here, ω_0 denotes the angular resonance frequency, assuming no losses. To see the behavior of the stored energy as a function of time it is possible to derive a differential equation from (2.1),

$$\frac{dU}{dt} = -\frac{\omega_0}{Q}U \tag{2.2}$$

with the solution $U(t) = U_0 e^{-\omega_0 t/Q}$. The initial amount of energy U_0 decays exponentially with a decay constant of $1/Q$[1]. The oscillations of the field inside the cavity can simply be written as

$$E(t) = E_0 e^{-\omega_0 t/2Q} e^{-i(\omega_0+\Delta\omega)t} \tag{2.3}$$

where $\Delta\omega$ is a possible shift of the resonance frequency. A damped oscillation consists of a superposition of frequencies around $\omega = \omega_0 + \Delta\omega$. Thus,

$$E(t) = \frac{1}{2\pi}\int_0^\infty E(\omega)e^{-i\omega t}d\omega \tag{2.4}$$

where

$$E(\omega) = \frac{1}{2\pi}\int_{-\infty}^\infty E_0 e^{-\omega_0 t/2Q} e^{i(\omega-\omega_0-\Delta\omega)t}dt\,. \tag{2.5}$$

Integrating (2.5) leads to the frequency distribution of the energy in the cavity

$$|E(\omega)|^2 \propto \frac{1}{(\omega-\omega_o-\Delta\omega)^2+(\omega_0/2Q)^2}\,. \tag{2.6}$$

The resonant line shape is a Lorentzian with a full width at half maximum $\delta\omega$ equal to ω_o/Q. The Q of a cavity is therefore

$$Q = \frac{\omega_0}{\delta\omega} = \omega_0\tau \tag{2.7}$$

where τ is the lifetime of the cavity and resonance, respectively. The cavity lifetime is connected with the linewidth of the resonance in angular frequency $\delta\omega$ simply by $\tau = \frac{1}{\delta\omega}$, as in the case of spontaneous emission of atoms [KS99].

In real cavities, the Q-factor is determined by several factors. For this reason the observed linewidth is a sum of all contributions from different loss mechanisms. The Q-factor of a given mode is therefore

$$\frac{1}{Q} = \sum\frac{1}{Q_i} \tag{2.8}$$

where the Q_i are the Q-factors associated to the various loss meachanisms. In the case of a microsphere there are, for example, material absorption, scattering on surface defects as well as on defects inside the glass, diffraction losses, pollution of the surface and coupling losses (see subsection 2.4). Usually one of these loss mechanisms dominates the others. For silica microspheres, Q-factors of up to 8×10^9 have been reported

[1]Measuring the intensity decay leads directly to the Q-factor of a cavity, exploited in cavity ring-down spectroscopy [SPOS97].

[GSI96]. To get an intuitive idea of the extremely high Q-factor of silica microspheres, one can consider a mechanical analogon, e.g. a tuning fork with a resonance frequency of 440 Hz (standard pitch). If such a tuning fork had a Q as high as a microsphere resonator with $Q = 8 \times 10^9$ it would vibrate for 33 days.
The Q-factor is normally used to characterize the resonator properties of silica microspheres. However, a high Q can always be obtained by increasing the cavity length, since the Q-factor is a measure for the photon storage time. The peculiarity of a microsphere resonator is the combination of the high Q-factor with a small mode volume (see section 2.3.4). A parameter for characterizing an optical cavity taking both the photon storage time which is directly connected with the cavity linewidth and the cavity size into account, is the so-called *Finesse*. It relates the **F**ree **S**pectral **R**ange (FSR) to the resonance linewidth by [Sie86]

$$\mathcal{F} = \frac{FSR}{\delta\nu} = 2\pi Q \frac{FSR}{\omega_0} \tag{2.9}$$

where

$$\delta\nu = \delta\omega/2\pi = \frac{c}{\pi L \sqrt{\frac{4R}{(1-R)^2}}} \tag{2.10}$$

is the width of the resonance. Here, L denotes the cavity length and R the reflectivity of cavity mirrors. The FSR represents the spacing between two longitudinal modes and is given by

$$FSR = \frac{c}{2LN} \tag{2.11}$$

where c is the speed of light and N the refractive index of the medium inside the cavity. The smaller the cavity, the larger the FSR. Inserting (2.10) and (2.11) into equation (2.9) one obtains an expression for $\mathcal{F}$ only depending on the reflectivity R

$$\mathcal{F} = \frac{\pi}{2} \frac{4R}{(1-R)^2} \,. \tag{2.12}$$

2.2 Analytic solutions for the fields inside a dielectric sphere

2.2.1 Vector wave equation

The starting point to derive the fields inside a dielectric sphere are the Maxwell equations [Jac98]

$$\nabla \cdot \mathbf{B} = 0 \tag{2.13}$$

$$\nabla \cdot \mathbf{D} = \rho \tag{2.14}$$

$$\nabla \times \mathbf{H} - \frac{\partial \mathbf{D}}{\partial t} = \mathbf{J} \tag{2.15}$$

$$\nabla \times \mathbf{E} + \frac{\partial \mathbf{B}}{\partial t} = 0 \tag{2.16}$$

where $\mathbf{B} = \mu \mathbf{H}$, $\mathbf{D} = \epsilon \mathbf{E}$ and $\mathbf{J} = \sigma \mathbf{E}$. They describe the behavior of the electric and magnetic fields in materials characterized by dielectric permittivity ϵ, magnetic permeability μ, and conductivity σ. However, within any closed domain of a homogeneous, isotropic medium without sources ($\rho = 0$ everywhere) the problem simplifies such that the field vectors **E**, **B**, **D**, and **H** satisfy one and the same partial differential equation. Let **C** denote any such vector, then

$$\nabla^2 \mathbf{C} - \mu\epsilon \frac{\partial^2 \mathbf{C}}{\partial t^2} - \mu\sigma \frac{\partial \mathbf{C}}{\partial t} = 0. \tag{2.17}$$

Without loss of generality one can make the assumption that the vector **C** contains the time only as a factor $e^{-i\omega t}$, where ω is the angular frequency. This allows a further simplification of the partial differential equation.

$$\nabla^2 \mathbf{C} + k^2 \mathbf{C} = 0 \tag{2.18}$$

where $k^2 = \epsilon\mu\omega^2 + i\sigma\mu\omega$. The strategy to solve this equation is to first solve the scalar version of (2.18), the so-called Helmholz equation [Jac98, Str99]

$$\nabla^2 \psi + k^2 \psi = 0 \tag{2.19}$$

and then to construct three independent vector solutions of (2.18) as follows:

$$\mathbf{L} = \nabla \psi, \tag{2.20}$$

$$\mathbf{M} = \nabla \times (\mathbf{a}\psi), \tag{2.21}$$

$$\mathbf{N} = \frac{1}{k} \nabla \times \mathbf{M} \tag{2.22}$$

where **a** can be any vector of unit length. It should be noted that the particular solutions of (2.19) which are finite, continuous, and single-valued in a given domain form a discrete set. Associated with each characteristic function ψ_n are the three vector solutions $\mathbf{L}_n$, $\mathbf{M}_n$, $\mathbf{N}_n$, no two of which are colinear. Thus, these characteristic vector functions allow to express any arbitrary wave function as a linear combination of them. If the divergence of the function vanishes, the expansion is made in terms of $\mathbf{M}_n$ and $\mathbf{N}_n$ alone. From the properties of $\mathbf{M}_n$ and $\mathbf{N}_n$ one can deduce the suitability for the representation of the fields **E** and **H**. Therefore one can write the general electric and magnetic field solutions as:

$$\mathbf{E} = -\sum_n (a_n \mathbf{M}_n + b_n \mathbf{N}_n) \qquad \mathbf{H} = -\frac{k}{i\omega\mu} \sum_n (a_n \mathbf{N}_n + b_n \mathbf{M}_n) \tag{2.23}$$

2.2.2 Scalar wave equation in spherical coordinates

After giving a general solution to the wave equation, it is time to turn to the specific problem of a sphere. According to the previous section one has to find a scalar function

$$\psi = f(R, \theta, \varphi)e^{-i\omega t} \tag{2.24}$$

where $f(R, \theta, \varphi)$ must satisfy equation (2.19) in spherical coordinates, defined as shown in figure 2.1:

$$\frac{1}{R^2}\frac{\partial}{\partial R}(R^2\frac{\partial f}{\partial R}) + \frac{1}{R^2 \sin\theta}\frac{\partial}{\partial \theta}(\sin\theta\frac{\partial f}{\partial \theta}) + \frac{1}{R^2 \sin^2\theta}\frac{\partial^2 f}{\partial \phi^2} + k^2 f = 0 \tag{2.25}$$

This equation is separable. When using the ansatz $f = f_1(R)f_2(\theta)f_3(\phi)$ one finds

$$R^2\frac{d^2 f_1}{dR^2} + 2R\frac{df_1}{dR} + (k^2R^2 - p^2)f_1 = 0 \tag{2.26}$$

$$\frac{1}{\sin\theta}\frac{d}{d\theta}(\sin\theta\frac{df_2}{d\theta}) + (p^2 - \frac{q^2}{\sin^2\theta})f_2 = 0 \tag{2.27}$$

$$\frac{d^2 f_3}{d\phi^2} + q^2 f_3 = 0 \tag{2.28}$$

The parameters p and q are separation constants. They need to fulfill the requirement that at any fixed point in space the field must be single-valued. q is restricted to the integers $m = 0, \pm 1, \pm 2, ...$, because f_3 is a periodic function with period 2π. The linearly independent solutions of (2.28) are therefore

$$f_{3,e} = \cos m\phi \qquad f_{3,o} = \sin m\phi \tag{2.29}$$

where the subscripts e and o denote even and odd functions. The solutions of f_2 after substitution of $\eta = \cos\theta$ in equation (2.27), are identified to be the associated Legendre polynomials of degree l and order m:

$$f_2(\eta) = P_l^m(\eta) = \frac{(1-\eta^2)^{\frac{m}{2}}}{2^l l!}\frac{d^{l+m}(\eta^2-1)^l}{d\eta^{l+m}} \tag{2.30}$$

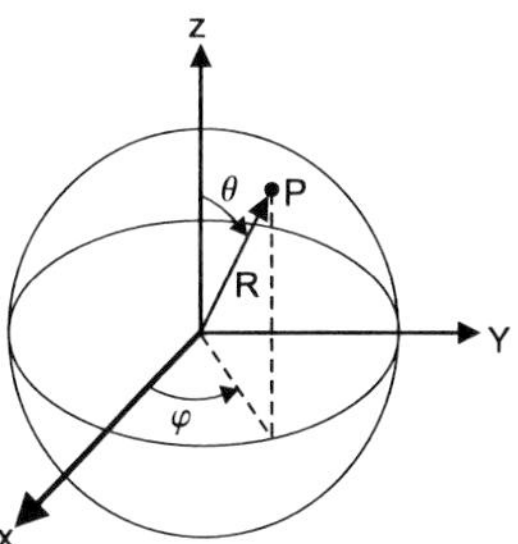

Figure 2.1: *Spherical coordinates.*

where p^2 was set to be equal $l(l+1)$ and $l = 0, 1, 2, \ldots$. It is worth noting that $P_l^m(\eta)$ vanishes for $m > l$ and m is restricted to the positive integers and zero[2]. In the last step the radial function $f_1(r)$ satisfying equation (2.26) needs to be identified. The solutions are the spherical Bessel functions:

$$f_1(r) = z_l(kR) \tag{2.31}$$

[2] The meaning of a negative m is shown later.

where $z_l(kR)$ can represent the spherical Bessel functions of the first and second kind $j_l(kR)$ and $n_l(kR)$, respectively, as well as the spherical Hankel functions $h_l^1(kR)$ and $h_l^2(kR)$ which are two special linearly independent combinations of $j_l(kR)$ and $n_l(kR)$. Details about the Bessel functions, their behavior and their generation by recurrence relations can be found in [AS72, Wat22]. Finally, it is possible to write down the generating functions that satisfy the scalar wave equation (2.19) in spherical coordinates.

$$f_{eml} = z_l(kR)P_l^m(\cos\theta)\cos m\phi \tag{2.32}$$

$$f_{oml} = z_l(kR)P_l^m(\cos\theta)\sin m\phi \tag{2.33}$$

The completeness of the functions $\cos m\phi$, $\sin m\phi$, $P_l^m(\cos\theta)$, $z_l(kR)$ allows to expand any function that satisfies (2.19) in spherical coordinates as an infinite series in terms of the functions (2.32) and (2.33). For convergence reasons, one chooses a Bessel function of the first kind for $z_l(kR)$ within domains which include the origin and a function of $h_l^1(kR)$ wherever the field is to be presented as a travelling wave.

2.2.3 Solutions of the vector wave equation in spherical coordinates

The solutions of the vector wave equation in spherical coordinates can be directly deduced from the generating functions (2.32) and (2.33). Putting the generating functions in (2.21) and (2.22) leads to the vector wave functions **M** and **N** the so-called *vector spherical harmonics.*

$$\mathbf{M}_{eml} = -\frac{m}{\sin\theta} z_l(kR)P_l^m(\cos\theta)\sin m\phi\hat{\mathbf{e}}_\theta - z_l(kR)\frac{\partial P_l^m}{\partial\theta}\cos m\phi\hat{\mathbf{e}}_\phi \tag{2.34}$$

$$\mathbf{M}_{oml} = \frac{m}{\sin\theta} z_l(kR)P_l^m(\cos\theta)\cos m\phi\hat{\mathbf{e}}_\theta - z_l(kR)\frac{\partial P_l^m}{\partial\theta}\sin m\phi\hat{\mathbf{e}}_\phi \tag{2.35}$$

$$\mathbf{N}_{eml} = \frac{l(l+1)}{kR} z_l(kR)P_l^m(\cos\theta)\cos m\phi\hat{\mathbf{e}}_R + \frac{1}{kR}\frac{\partial}{\partial R}[Rz_l(kR)]\frac{\partial}{\partial\theta}P_l^m(\cos\theta)\cos m\phi\hat{\mathbf{e}}_\theta - \frac{m}{kR\sin\theta}\frac{\partial}{\partial R}[Rz_l(kR)]P_l^m(\cos\theta)\sin m\phi\hat{\mathbf{e}}_\phi \tag{2.36}$$

$$\mathbf{N}_{oml} = \frac{l(l+1)}{kR} z_l(kR) P_l^m(\cos\theta)\sin m\phi \hat{\mathbf{e}}_R +$$
$$\frac{1}{kR}\frac{\partial}{\partial R}[Rz_l(kR)]\frac{\partial}{\partial\theta}P_l^m(\cos\theta)\sin m\phi\hat{\mathbf{e}}_\theta + \tag{2.37}$$
$$\frac{m}{kR\sin\theta}\frac{\partial}{\partial R}[Rz_l(kR)]P_l^m(\cos\theta)\cos m\phi\hat{\mathbf{e}}_\phi$$

In order for $\mathbf{M}_{oml}$ to be tangential over the entire surface of the sphere, the vector $\mathbf{a}$ in equation (2.21) needs to be parallel to the radial vector R. According to subsection 2.2.1, any solution of the vector wave equation in spherical coordinates can be expressed as a sum of these vector spherical harmonics. The subscript n in equation 2.23 now stands for all indices, which are three in this case. The time dependency can easily be included with the factor $e^{-i\omega t}$.

2.2.4 Natural modes of a sphere and their field distribution

Consider a sphere with radius a characterized by the constant $k_1 = \sqrt{\varepsilon_1\mu_1\omega^2 + i\sigma_1\mu_1\omega}$, which is embedded in a finite homogeneous medium of k_2 (see section 2.2.1). The field can be expressed with equation (2.23). If the coefficients a_n are all zero, and if only the b_n are excited, the field has a radial component of $\mathbf{E}$, but the magnetic vector is always perpendicular to the radius vector (see equations (2.23) and (2.34) – (2.37)). In a waveguide or cavity such modes are called transversal magnetic (TM). If only the a_n are excited, the electric field has no radial component and is thus labeled as transversal electric (TE). The coefficients in equation (2.23) are as yet set arbitrarily. Without loss of generality, it is possible to switch from the even and odd solutions to a description where the ϕ-dependency is expressed in terms of $e^{\pm im\phi}$, denoted by $+$ and $-$. The difference between the two solutions is now the round trip direction. Let's consider a TE-mode with an $e^{im\phi}$ dependency.

$$\mathbf{E} = -a_{+ml}\mathbf{M}_{+ml} \qquad \mathbf{H} = -\frac{k}{i\omega\mu}a_{+ml}\mathbf{N}_{+ml} \tag{2.38}$$

The functions $\mathbf{M}_{+ml}$ and $\mathbf{N}_{+ml}$ must be chosen such that the field is finite at the origin and regular at infinity. Therefore the Bessel function of the first kind is an appropriate solution. For $R < a$ the electric field has the following form[3]:

$$\mathbf{E}^{in}_{+ml}(R,\theta,\phi) = -ima^{in}_{+ml}\frac{1}{\sin\theta}P_l^m(\cos\theta)j_l(k_1R)e^{(im\phi)}\hat{\mathbf{e}}_\theta -$$
$$a^{in}_{+ml}j_l(k_1R)e^{(im\phi)}\frac{\partial P_l^m(\cos\theta)}{\partial\theta}\hat{\mathbf{e}}_\phi \tag{2.39}$$

The external field in the region $R > a$ can be obtained by replacing the Bessel function $j_l(kR)$ by $h_l^1(kR)$ and substitution of k_1 and a^{in}_{+ml} accordingly.

[3]The explicit solution for $\mathbf{H}_{+ml}(R,\theta,\phi)$ can be derived in the same way.

$$\mathbf{E}^{out}_{+ml}(R,\theta,\phi) = -ima^{out}_{+ml}\frac{1}{\sin\theta}P_l^m(\cos\theta)h_l^1(k_2R)e^{(im\phi)}\hat{\mathbf{e}}_\theta -$$

$$a^{out}_{+ml}h_l^1(k_2R)e^{(im\phi)}\frac{\partial P_l^m(\cos\theta)}{\partial\theta}\hat{\mathbf{e}}_\phi \tag{2.40}$$

For large l, $h_l^1(kR)$ can be approximated by [GI94]

$$\exp(ikd) \tag{2.41}$$

where $k \approx 2\pi/\lambda\sqrt{1-N^2}$, d is the distance to the sphere's surface, N the refractive index of the sphere and λ the wavelength of light. It is further assumed that the refractive index of the surrounding medium is 1. Since N is bigger than 1, k is complex. Therefore this field component is not propagating, but exponentially decaying. Such a field is called *evanescent*. Equation (2.41) can be rewritten as

$$\exp(-d/r^*) \tag{2.42}$$

where $r^* \approx \lambda/2\pi \times (\sqrt{N^2-1})$ is the decay length of the evanescent field. This field outside the sphere gives access to the mode inside the sphere in a natural way.
With the explicit field solutions it is finally possible to determine the resonance frequencies of the sphere by solving a boundary value problem. The boundary conditions are given by the continuity of the tangential components of $\mathbf{E}$ and $\mathbf{H}$ at the surface. They are fulfilled by a discrete set of characteristic values $\rho = k_2a$, which are the roots of a transcendental equation,

$$\frac{[N\rho j_l(N\rho)]'}{\mu_1 j_l(N\rho)} = \frac{[\rho h_l^1(\rho)]'}{\mu_2 h_l^1(\rho)} \tag{2.43}$$

where the prime denotes the derivative. The roots are labeled by a number $n = 1, 2, 3, ...$ A set of natural frequencies corresponds to the allowed set of values ρ. These are known as the modes of oscillation, where $n = 1$ is associated with the lowest frequency. Since there is no dependency in m, the solutions are degenerate in m for a perfect sphere. The field distribution and the transcendental equation for the TM-modes can be found in exactly the same manner.

2.3 Whispering-gallery modes

The last section provided analytic solutions for the field distribution in a dielectric sphere. It was shown that the occurring resonances have different polarizations (TE and TM modes) and are labeled by three mode numbers n, l, m, like atomic bound states. But from all the modes present in a sphere only a special subclass is of interest for this work, the so-called whispering-gallery modes with their high-Q factor. However, the vector spherical harmonics do not give a direct access to the whispering-gallery modes in a sense which allows to distinguish them from other modes. Also they do not provide a simple interpretation of the mode numbers. Approaches based on approximations help to get a deeper and more intuitive understanding of these particular modes.

2.3.1 Radial intensity distribution

In order to introduce the concept of whispering-gallery modes in a heuristic way, some geometrical optics is used. Consider a microsphere with radius a and refractive index N. A ray of light with wavelength λ is propagating inside, hitting the surface under an angle α_{in}. If $\alpha_{in} > \alpha_{critical} = \arcsin(\frac{1}{N})$, the light is totally internally reflected. Because of the symmetry of the sphere, all subsequent reflections occur under the same angle, and the light is trapped as a result (figure 2.2 a)). This simple geometrical

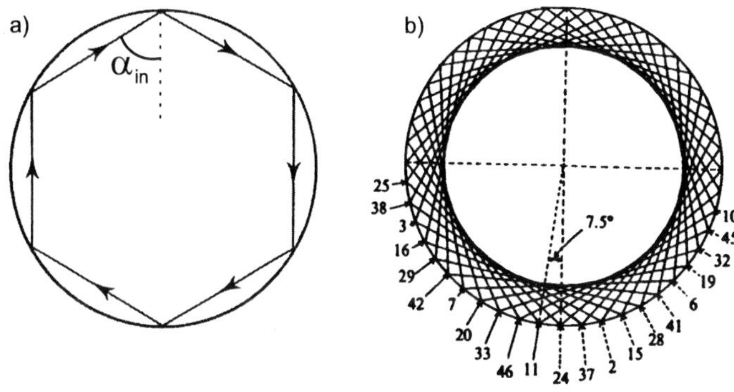

Figure 2.2: *a) A resonance in a sphere: the light ray is trapped inside self consistently. b) The light ray superimposes only after* 11 *round tips. On its way it was reflected* 48 *times in all [Col94].*

picture leads immediately to the concept of resonances [CC96]: For large spheres where $\lambda << a$ the ray propagates close to the sphere's surface, and traverses approximately $2\pi a$. The resonance condition is fulfilled, if a round trip equals an integer number l of wavelengths leading to constructive interference and the built-up of a mode. In other words the phase needs to match after one round rip. The modes where the light is trapped, travelling in a great circle around the sphere's perimeter, are the so-called *whispering-gallery modes.* l is one of the mode numbers already discussed in section 2.2, here as a tangible number for a resonance condition

$$\lambda \approx \frac{2\pi a N}{l}. \tag{2.44}$$

For these modes the light is strongly confined close to the surface. For slightly smaller angles of incidence, the light ray needs two round trips for a self consistent superposition. The mode is penetrating deeper inside the sphere [Rol99], thus creating another set of resonances with a different wavelength, but with the same l. Figure 2.2 b) is an example of a mode which needs 11 round trips to superimpose. It turns out that this radial dependency of the field distribution depends on the mode number n. In figure 2.3 the angle averaged intensity distribution is plotted for three modes, only differing in n. The bigger the n the more the strongest intensity peak moves towards the center of the sphere. The radial extent of the mode depending on n and l can be approximated by [Sch93]

$$\frac{\Delta R}{a} \approx 2.2[(n - 1/4)/Nx_{l,n}]^{2/3} \tag{2.45}$$

where $x_{l,n} = (2\pi/\lambda)a$ is the sphere's *size parameter* which relates the dimensions of the cavity to the wavelength used. n is sometimes called the radial mode number, due to the radial dependency of the field upon it. n is the number of intensity maxima inside the sphere or, in other words, $n-1$ is the number of zeros of the Bessel function $j_l(kR)$ inside the resonator. It also becomes apparent that the higher n is, the more the field leaks out of the sphere.

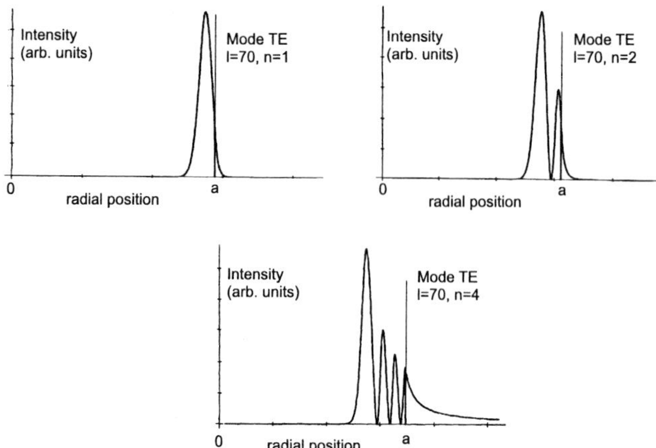

Figure 2.3: *Radial intensity distribution for three modes differing in n [Col94].*

2.3.2 Angular intensity distribution

The dependency on ϕ can immediately be seen by having a closer look at equations (2.34) – (2.37). In ϕ direction the field depends only on $\sin m\phi$ or $\cos m\phi$ and $e^{-im\phi}$ or $e^{+im\phi}$, respectively. The number of intensity peaks in this angular direction is therefore $2m$, demonstrating again the simple meaning of the mode numbers. In figure 2.4 the intensity distribution in the equatorial plane is shown for a TE whispering-gallery mode with $n = 1$ and $m = l = 9$. The radius of the sphere is normalized to 1. The graph is

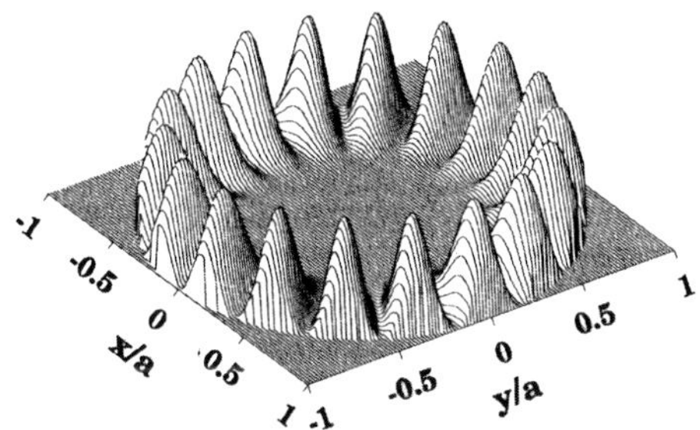

Figure 2.4: *Intensity distribution in the equatorial plane for a TE whispering-gallery mode with $n = 1$ and $m = l = 9$ [CC96]*

based on computer codes for spherical wave functions, which can be found in reference [BH90].
The intensity distribution in θ direction is not obtainable in a straightforward manner. Let's start with the simplest case where $l = m$ and l is large. Then $P_l^l(\cos\theta)$ can be approximated by [BS91]

$$\frac{P_l^l(\cos\theta)}{1 \cdot 3 \cdot 5 \cdot \ldots \cdot (2l-1)} = \sin^l\theta \approx e^{(\frac{-l(\theta-\frac{\pi}{2})^2}{2})}. \tag{2.46}$$

Such a mode can be interpreted as a "Gaussian" beam having l/n reflections on the surface during one evolution inside the sphere [GI94]. A simple calculation about the width of that Gaussian beam leads to an approximation of the extent of the mode in the θ direction,

$$D \approx a\sqrt{\frac{8}{l}} \tag{2.47}$$

where D is the mode's diameter. For a perfect sphere, that Gaussian beam can be inclined at different angles.
However, this idealized picture does not hold for high-Q silica microspheres, because due to their fabrication method they are actually never perfectly spherical. As a result the symmetry is broken, an equatorial plane is defined and the m degeneracy is lifted. The induced small ellipticity makes the interpretation of the associated Legendre polynomials $P_l^m(\cos\theta)$ less evident. With the growth of the difference between l and m, $P_l^m(\cos\theta)$ becomes a quickly oscillating function with a sharp cutoff at

$$\theta_{max} = \frac{\pi}{2} \pm \arccos(\frac{m}{l}) \tag{2.48}$$

An approximation can help reveal the nature of the θ dependency. For large l and $|m| \approx l$ the intensity distribution on the entire sphere surface can be approximated by

[KDS+95]:

$$I_{l,m}(\theta,\phi) \propto |H_{l-|m|}(l^{\frac{1}{2}}\cos\theta)\sin^{m}(\theta)\exp(im\phi)|^{2} \qquad (2.49)$$

where $H_{l-|m|}$ represents a Hermite polynomial [GR00]. Figure 2.5 shows plots of $I_{l,m}(\theta)$ for 4 different m values. l was set to be 350, which is a realistic value for silica microspheres studied in this work. A mode characterized by n, l, m has $l-|m|+1$

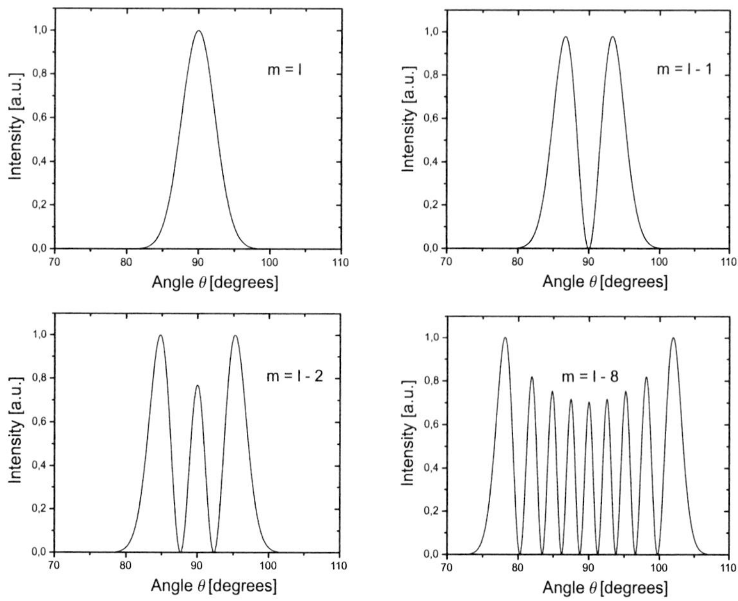

Figure 2.5: *Intensity distribution along the θ-direction for different m values.*

lobes in the θ direction, centered around $\pi/2$ the equatorial plane of the sphere. The most confined mode with $l=m$ is called a *fundamental mode* of the microsphere. From figure 2.5 one can also see that with increasing difference between m and l the modes become more extended, as described by (2.48), while at the same time the distance h between adjacent intensity maxima becomes smaller. From geometrical considerations one can write

$$\sin\alpha \approx \frac{h}{a} \qquad (2.50)$$

where α is the angle between the lines joining the two intensity maxima to the spheres center. Together with the relation [GI94]

$$\cos\alpha \approx \frac{m}{l} \qquad (2.51)$$

h can be calculated. This last relation completes the discussion about the intensity distribution of a whispering-gallery mode on the surface of a sphere.

2.3.3 Spectral properties of whispering-gallery modes

The spectral positions of the resonances are, in principle, already given by the solutions of equation (2.43). In order to avoid the necessity to solve this transcendental equation with high order Bessel functions explicitly, a systematic asymptotic analysis of equation (2.43) can be performed ([SB91, LLY92]). Asymptotic series expansions were found, leading to a rather simple expression for the spectral positions of the resonances

$$Nx_{l,n} = l + \frac{1}{2} - \alpha_n \left(\frac{l+1/2}{2}\right)^{1/3} - \frac{Np}{\sqrt{N^2-1}} + \frac{3\alpha_n^2}{2^{2/3}10(l+1/2)^{1/3}} - \frac{N^3 p(p^2/3-1)\alpha_n}{2^{1/3}(N^2-1)^{3/2}(l+1/2)^{2/3}} \tag{2.52}$$

where $x_{n,l}$ again denotes the size parameter (see equation (2.45)). For TM modes this parameter is written as $a_{n,l}$ and $p = 1/N^2$, while for TE modes $x_{n,l} = b_{n,l}$ and $p = 1$ [SB91]. α_n denotes the nth zero of the Airy function $Ai(\alpha)$ [AS72]. It was shown in [Sch93], that the accuracy of the series expansions can be better than 10^{-4} for the $n = 1$ mode with $l > 50$. The free spectral range (FSR) of a microsphere can be directly obtained from (2.52) by using the definition of the size parameter,

$$FSR = \frac{c}{2\pi a}[x_{n-1,l} - x_{n,l}] \approx \frac{c}{2\pi a N}. \tag{2.53}$$

One sees that this is the FSR of a wave propagating along the inner surface of the sphere by repeated total internal reflection. The total internal reflection process leads also to a frequency shift

$$c_{TE/TM} = \frac{a_{n,l} - b_{n,l}}{FSR} \approx \frac{\sqrt{N^2-1}}{N} \tag{2.54}$$

between the TE and TM spectra, because the TE and TM modes experience different phase shifts in every internal reflection along one round trip [SB91]. $c_{TE/TM}$ is a constant for a given N, which relates the positions of TE and TM modes with the same n and l. The TM mode is the mode of higher frequency and lower wavelength, respectively. The series expansion (equation (2.52)) allows to calculate the spectral positions of TE and TM modes depending on n and l in a rather simple way, but it cannot predict the positions of the modes differing in m because an ideal sphere is assumed.

The problem of determining the position of modes differing only in m is tackled in reference [GI94] by means of precessing modes in a slightly deformed silica microsphere. There it is shown, that an arbitrary spherical function can be modelled by the precession of an inclined fundamental mode. The frequency shift can then be interpreted as a change in the perimeter length of the inclined ellipse of elipticity ϵ. A first order approximation yields

$$\frac{\Delta\omega}{\omega} = \pm\frac{\epsilon^2(l^2-m^2)}{4l^2} \tag{2.55}$$

where $\Delta\omega$ denotes the distance of a mode with certain m in frequency space to the fundamental mode of the same family. A family of modes is built by all modes with the same n and l. The lift of the degeneracy can be towards higher or lower frequencies depending on the shape of the sphere. The positive sign in equation (2.55) stands for an oblate spheroid, while the negative sign stands for a stretched one.

2.3.4 Mode volume of a whispering-gallery mode

The knowledge of the electric field distribution of a mode allows to calculate the mode volume. Due to the inconvenient form of the the vector spherical harmonics, an approximation is used once again [BGI89, GPI00]:

$$V_{mode} \approx \frac{(\int \mathbf{E}^2 d^3r)^2}{\int \mathbf{E}^2\mathbf{E}^2 d^3r} \approx 3.4\pi^{3/2}(\lambda/2\pi N)^3 l^{11/6}\sqrt{l-m+1} \tag{2.56}$$

This expression holds only for modes with $n = 1$. As an example of how big the mode volume of a mode present in a silica microsphere (N=1.46) is, let's assume $n = 1$, $l = 300$, $m = l$ and $\lambda = 670\,nm$. The resulting mode volume is $250\,\mu m^3$ for a sphere with a diameter of $44\,\mu m$, while its total volume is $44602\,\mu m^3$.

2.4 Efficient coupling to whispering-gallery modes

In section 2.3.1 the simple picture from geometrical optics was used to introduce the concept of whispering-gallery modes. There, a light ray is trapped inside the sphere by repeated total internal reflection. The reciprocity theorem, however, tells in this case that if no light is coupled out of the resonator, there is no trivial way to couple light into it [4]. A solution for efficient coupling is offered by the evanescent field outside the sphere (see subsection 2.2.4). To couple light efficiently into the sphere the phase matching condition has to be met: the k vector of the incoming light needs to be matched to the k vector of the light inside the whispering-gallery modes[5]. All experimentally demonstrated methods therefore use evanescent coupling: there are prism couplers [BGI89], fiber tapers [KCJB97], eroded monomode fibers [DKL$^+$95], angle polished single mode fibers [IYM99] and pedestal antiresonant reflecting waveguides [LLL$^+$00]. Although each coupling method has its own advantage, the prism coupler is by far the most common. An evanescent field is created, if one launches the light under an angle Ψ bigger than $\sin\Psi_{tot} = \frac{1}{N_{prism}}$, the critical angle for total internal reflection on the inner surface of a prism (see figure 2.6). The k vector in the air close to the interface were the light beam is internally reflected can be decomposed in a $k^{\perp}$ which is perpendicular to the interface and a $k^{\parallel}$ which is parallel to the prism surface [CMD72]:

$$k^{\perp} = \frac{2\pi}{\lambda}(1-(N_{prism}\sin\Psi)^2)^{1/2} \tag{2.57}$$

[4]By using standard Mie theory one can show, for example, that a plane wave cannot excite a whispering-gallery mode.

[5]Phase matching reflects the conservation of photon momentum.

$$k^{\|} = \frac{2\pi}{\lambda}(N_{prism} \sin \Psi). \tag{2.58}$$

For $\Psi > \Psi_{tot}$ the perpendicular component of the k vector is purely imaginary (see equation (2.57)). Therefore, the electrical field $\mathbf{E}(\mathbf{r}) \propto e^{i(\mathbf{kr}-\omega t)}$ has a real exponent,

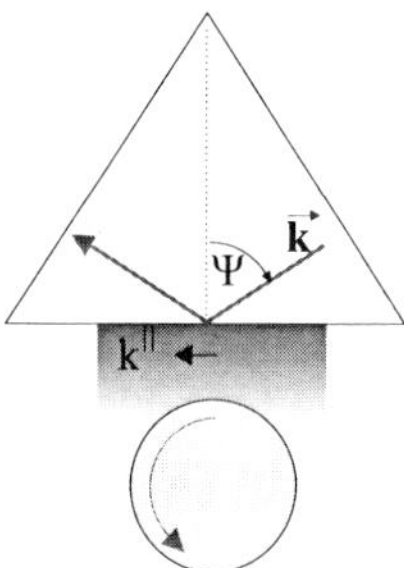

Figure 2.6: *Efficient coupling to a microsphere via the evanescent field of a prism coupler.*

it is not propagating anymore in the direction perpendicular to the prism surface, but exponentially decaying with increasing distance to the prism surface. In order to couple light into a microsphere the parallel component of the k vector has to be matched to the parallel component of the evanescent field of the microsphere:

$$\frac{2\pi}{\lambda} N_{prism} \sin \Psi = \frac{2\pi}{\lambda} N_{sphere} \sin \Psi_{sphere} \,. \tag{2.59}$$

The angle inside the sphere is approximately $\Psi_{sphere} = 90°$ for the high-Q modes. Then one obtains as a condition for the launching angle Ψ on the prism

$$\sin \Psi = \frac{N_{sphere}}{N_{prism}} \,. \tag{2.60}$$

Thus, the refractive index of the prism has to be bigger than that of the sphere. The refractive index of the silica microspheres is $N_{sphere} = 1.46$ and therefore a prism made out of a high refractive index glass was chosen with $N_{prism} = 1.72$. This leads to an optimal coupling angle of $\Psi = 59°$. The inclination angle with respect to the equatorial plane was not yet considered, but for symmetry reasons it should be $0°$. In references [GI94] and [GI99], systematic theoretical and experimental investigations were performed to optimize the coupling conditions to high-Q whispering-gallery modes. It was shown that modes differing in m have different optimal launching angles, but in the case of the fundamental mode $l = m$ the equation for Ψ is identical to equation (2.60) and the angle of incidence with respect to the equatorial plane is also $0°$.
So far, only the prerequisites on the incoming light on the prism were considered, but for efficient coupling there exists also an optimum distance between the sphere and

the prism for the following reason: the output signal of the prism when the sphere is coupled can be considered as the result of interference between the input light and the reemission of the resonator [GI99]. If the sphere is too close to the prism, not only the coupling of light into the sphere but also its extraction out of the sphere is large. In the ideal case, the output intensity at the prism is zero, i.e. the entire input power is lost inside the resonator (critical coupling) [CPV00]. The sphere-prism gap is usually experimentally adjusted (to about a distance of the decay length of the evanescent field), such that the coupling is strongest (see chapter 4).

2.5 Table with important quantities

The last section of this chapter summarizes the important quantities introduced in the previous sections. It is not only a reference table, but it helps to form an impression about the dimensions of the quantities, before entering the experimental part.

Symbol	**explanation**	**typical values**
a	radius of the sphere	$20 - 150\,\mu m$
n	radial mode number: number of intensity maxima in the radial direction inside the sphere	1-2
l	angular mode number: number of wavelengths fitting along the spheres circumference ($l \approx 2\pi a N/\lambda$)	300-2000
m	angular mode number: 2m intensity peaks in the ϕ (azimuthal) direction	300-2000
$l - \|m\| + 1$	number of intensity maxima perpendicular to the sphere's equator	1 for the fundamental mode
V_{mode}	mode volume of a mode with $n = 1$	$250 - 8000\,\mu m^3$
$\Delta\omega$	denotes the distance of a mode with certain m in frequency space to the fundamental mode of the same family	$0.5 - 10\,GHz$ for the $l - \|m\| + 1 = 2$ mode
ϵ	ellipticity of the sphere	about $1\,\%$
$c_{TE/TM}$	constant which relates the positions of the TE and TM modes with the same n and l (TE/TM-splitting)	0.73, TM has the higher frequency
FSR	free spectral range	$200\,GHz - 1.6\,THz$
$\Delta\nu$	resonance linewidth	about $1\,MHz$
Q	Q-factor	several times 10^8, in some cases $> 10^9$
$\mathcal{F}$	Finesse	$> 10^5$, sometimes more than 1 million

Chapter 3

The tools

During this work several tools were developed in order to perform the experiments on high-Q microspheres. The main tools are described in this chapter. Some basics are given, as well as a detailed description of their implementation. However, to understand the chapters that follow it is not necessary to know all the technical details. If one is familiar with tunable grating stabilized diode lasers, with confocal and near-field microscopy, it is possible to skip this chapter and go to chapter 4, where the description of the experiments on microspheres starts.

3.1 The grating stabilized diode laser

For characterization of microsphere resonators it is implicitly necessary to perform spectroscopy in order to determine the spectral positions and widths of the resonances as well as the diameter and ellipticity of the sphere. The most important property of the sphere for this work is the Q-factor. Equation 2.7 shows two possible ways to measure it. One either determines the cavity's decay time (cavity ringdown spectroscopy) [VIM$^+$98] or one measures directly the width $\delta\nu$ of the resonance [BGI89]. A prerequisite for both types of measurements is a narrow linewidth tunable laser source. Especially for determining the width of a mode it is necessary that the laser linewidth is smaller than that of the mode to be measured. A Q-factor of 10^9 at a visible wavelength (e.g. $670\,nm$) corresponds to a linewidth of only $447\,kHz$. Therefore the laser linewidth should be on the order of $100\,kHz$ on a timescale necessary to sweep over the resonance. A relatively inexpensive, robust and easy to use laser, which can fulfill these requirements is a grating stabilized diode laser.

3.1.1 Operation principles

In a simplified picture, a diode laser consists of a p-n-junction with two plane parallel cleaved end facets, which have optical quality. Due to the high refractive index of the semiconductor material (e.g. $N = 3.6$ for GaAs), these facets have a reflectivity of more than $30\,\%$ and thus act as a Fabry-Perot resonator. A current running through the

semiconductor creates electron-hole pairs, which emit photons when they recombine. The cleaved end facets provide enough feedback to enable laser action. Such a diode laser has a gain profile of a few tens of nanometers depending on the material used. By changing the temperature or the current through the diode, the refractive index of the material changes and the maximum of the gain profile shifts, thus allowing to tune the wavelength[1]. However, in both cases it is not possible to change the optical path length and the gain profile synchronously [Wen94]. As a result not all wavelengths are accessible (tuning gaps) and during the wavelength tuning mode-hops tend to occur. Moreover, the linewidth of free-running diode lasers is generally on the order of 20 MHz [PW91], unacceptable for spectroscopy of high-Q microspheres.
Tunability and linewidth can be significantly improved by optical feedback to increase the effective Q-factor of the cavity, resulting in an increased photon storage time and thus in a linewidth narrowing. A quite frequently used technique is to use a grating, which forms an external cavity with the back facet of the laser diode. This arrangement is known as Littrow configuration (see figure 3.1). To calculate the optimal angle for

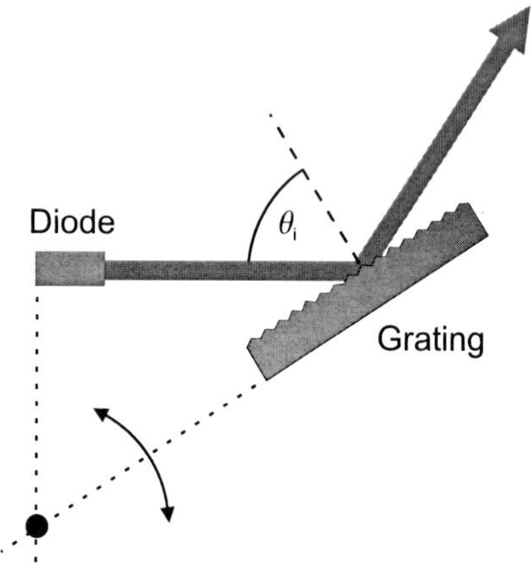

Figure 3.1: *Scheme of a grating stabilized diode laser in a Littrow configuration.*

the light on the grating one uses the grating equation

$$d(\sin\theta_i + \sin\theta_m) = m\lambda \tag{3.1}$$

where $\sin\theta_m$ is the diffraction angle of the m-th order and d is the grating constant. The angle of incidence θ_i on the grating is chosen such that the minus first order is diffracted back into the diode. The 0-th order is coupled out. Setting $\theta_{m=-1} = \theta_i$ leads

[1]For time scales longer than 1 μs a current tuning causes mainly a fast change in the temperature and is thus also a kind of temperature tuning [WH91].

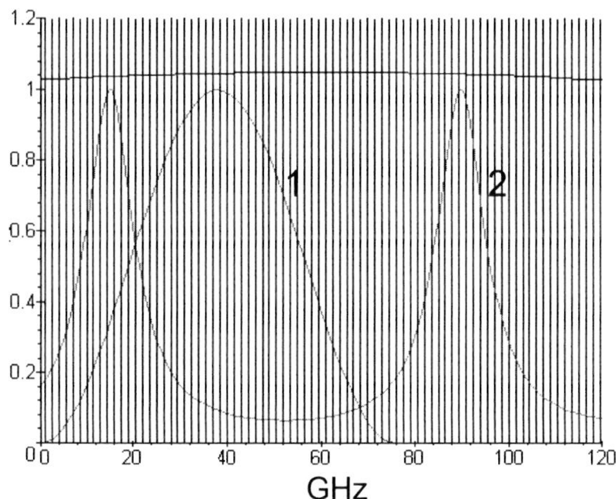

Figure 3.2: *Spectral properties of the components of a grating stabilized diode laser. The narrow lines denote the resonances of the external cavity with a FSR of 1.5 GHz. The upper line represents the gain profile of the semiconductor, which is broader than 1000 GHz. Curve 1 represents the dispersion of the grating ($FWHM = 38\,GHz$) and curve 2 the resonances of the laser diode itself with a FSR of 75 GHz and a Finesse of 6. These values hold for a 500 μm cavity and a refractive index of $N = 4$ [Dem99].*

to the equation for the angle of incidence

$$\theta_i = \arcsin \frac{\lambda}{2d}. \tag{3.2}$$

Changing the angle of the grating should lead to wavelength tuning. Unfortunately, equation (3.2) is not the only condition for tuning. In figure 3.2 the situation is sketched and the interplay of the different elements can be seen. In order to tune such a laser mode-hop free, the grating angle has to be changed synchronized with the length of the semiconductor cavity and the external cavity, which need to fulfill

$$L = n\lambda/2N \tag{3.3}$$

where L is the cavity length, N the refractive index inside the cavity and n the number of half wavelengths fitting into the cavity. It turns out that the conditions (3.2) and (3.3) are simultaneously fulfilled for the external cavity if the grating turns around an axis, which is determined by the intersection of two lines, one starting at the end facet of the laser diode, the other starting at the grating (see figure 3.1). The optical path

length in the semiconductor is adjusted by ramping the current during a scan, such that this cavity follows the frequency change, accordingly.
The tuning properties could be much improved if one could eliminate the internal cavity. Then one would only have to be concerned about the conditions (3.2) and (3.3) for the external cavity. This can be achieved with an antireflection coating on one output facet. A continuous tuning range of $15\,nm$ at a center wavelength of $1260\,nm$ was achieved with an antireflection coated diode and a grating providing the feedback [FlGSL86]. But not only the tuning characteristics are much better with such an antireflection coating, also the spectral linewidth of the laser gets much narrower. Values between $300\,Hz$ and $50\,kHz$ have been reported [Wya85, BBHS91], depending on many parameters, such as emission wavelength, feedback, residual reflectivity of the facet and others.

3.1.2 Setting up the diode laser

A sketch of the grating-stabilized diode laser can be seen in figure 3.3. As laser diode, an antireflection coated diode emitting around $672\,nm$ from Sacher Lasertechnik (SAL-675-10, residual reflectivity smaller than 10^{-4}) is used. It is mounted in a collimation tube with a high numerical aperture lens (NA=0.55) in order to collimate the strongly diverging beam. This tube can be fixed in an aluminium construction on a thermic insulated baseplate which is cooled with a Peltier element slightly below room temperature. A feedback loop stabilizes the temperature for minutes in the millikelvin range, while a long term stability of 10 millikelvin is achieved. The generated heat is transported away by a water cooled brass base. Optical feedback is provided by a grating ($1800\,grooves/mm$) diffracting about 20 % into the minus first order, directed back to the diode. The angle of incidence on the grating was chosen according to equation (3.2) to be 37°. This grating, mounted on a modified mirror mount, is placed at a distance of about $10\,cm$ to the laser diode. The length was chosen to compromise the photon lifetime in the cavity and the mechanical and thermal stability, respectively. A longer cavity increases the photon lifetime, but it also gets more sensitive to mechanical vibrations and thermal effects. Various precautions were taken to reduce mechanical vibrations and to guarantee a narrow linewidth. The tuning is done by two piezoelectric elements (hub $2\,\mu m$) inserted between the fine thread screws of the mirror mount and the plate holding the grating (see figure 3.3). The two piezoelectric elements are connected via a potential divider, to move the grating, as if it would move around the optimal turning point. This trick avoids the necessity to know the exact position of the turning point. By adjusting the potential divider one can experimentally determine a good relation between the travel of the two piezoelectric elements and achieve a continuous tuning. Additionally, the current can be ramped synchronously as an aid for tuning the cavity length. A major improvement concerning the tunability was the integration of a slit into the cavity, in order to avoid mode-hops of $1.5\,GHz$, corresponding to a jump of one FSR of the $10\,cm$ long cavity. This blackened slit cuts

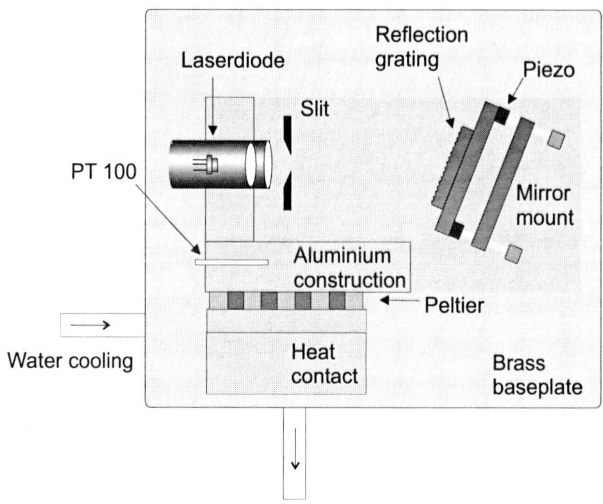

Figure 3.3: *Sketch of the grating stabilized diode laser [Dem99].*

about 1/3 of the laser beam's diameter, acting as an additional mode selective element.

3.1.3 Performance

In figure 3.4 the output power versus injection current (L/I characteristic) is shown. The diode delivers more than $7\,mW$ of output power still working in the linear regime, which means working with such a power does not have a negative influence on the lifetime of the diode. The laser could be scanned about $8\,GHz$ mode-hop free with the piezoelectric elements, monitored and gauged with a confocal Fabry-Perot ($\mathcal{F} = 150$, mode spacing $3\,GHz$). Varying the current between 40 and $60\,mA$ allowed to span $60\,GHz$ with several overlapping piezo element scans, which were put together with a wavemeter. By changing the grating angle or the temperature several hundred GHz could be covered, enough to investigate an entire FSR of a large $200\,\mu m$ microsphere, with a FSR of about $320\,GHz$.
The linewidth of the laser was determined with a self-heterodyne measurement [OKN80]. Experimental details are described in reference [Dem99]. Figure 3.5 shows the beat signal, from which the linewidth can be extracted by taking half of the FWHM of the Lorentzian fit [GD84]. A value of less than $4\,kHz$ is determined falling in the range reported for such lasers (see references mentioned above). However, on longer timescales (hundreds of milliseconds) the acoustic jitter of the cavity increases the linewidth to a few hundred kHz. This conjecture is supported by experiments with high-Q microspheres, where the scan speed of the laser was varied. With increasing

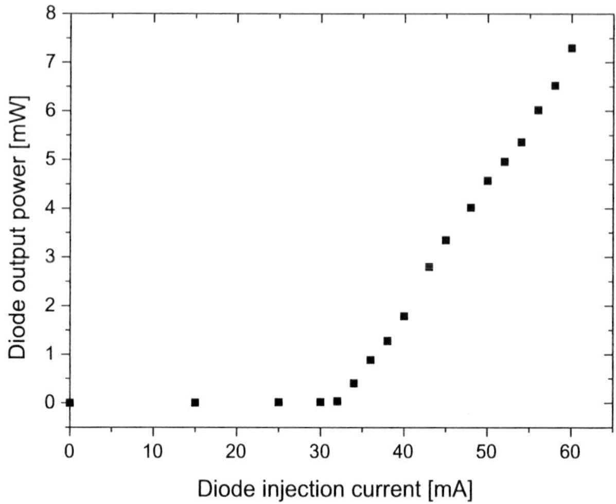

Figure 3.4: *L/I characteristic of the diode laser.*

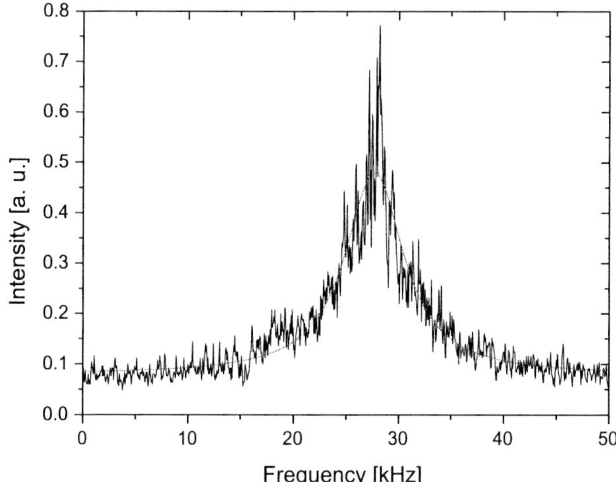

Figure 3.5: *Self-heterodyne beat spectrum. The width (FWHM) of the Lorentzian fit is less than* $8\,kHz$.

scan speed, the resonances got less "noisy" and narrower until a Lorentzian lineshape was to be observed. Using a scan speed of about 100 Hz, which was a limit given by the data acquisition card, used to control the diode laser and to record the data, allowed to measure linewidths of microsphere resonances of about 1 MHz (corresponding to a Q of several times 10^8). At the end of this work, a commercial diode laser (New Focus 6300 Velocity) was available, with a continuous tuning of 60 GHz and a linewidth of less than 300 kHz in a timescale of 50 ms allowing to measure Q factors $\geq 10^9$.

3.2 The beam scanning confocal microscope

In recent years, ultra sensitive microscopy was established as a powerful tool for single molecule or single quantum dot spectroscopy [Moe02, ZEK02, ENSB99]. Especially confocal techniques are versatile for exciting and detecting only a single quantum emitter at a time [SBN00, FSZ+00]. For this work, a microscope was needed with the ability to detect and to address a single nanoparticle on the surface of a microsphere, which is at the same time coupled to a prism coupler. This condition led to some particular requirements that had to be met, which could not be fulfilled by standard (commercially available) microscopes.

3.2.1 Confocal microscopy

The principle of confocal microscopy is sketched in figure 3.6. A point source (e.g. a laser focused on a pinhole) is imaged into the object plane of a microscope objective, such that the illuminated point on the sample is in a conjugated (=confocal) plane of the point source. Then the objective forms an image of the illuminated point on a pinhole. Point source, illuminated point on the sample and imaged point on the pinhole are mutually confocal - hence the name.

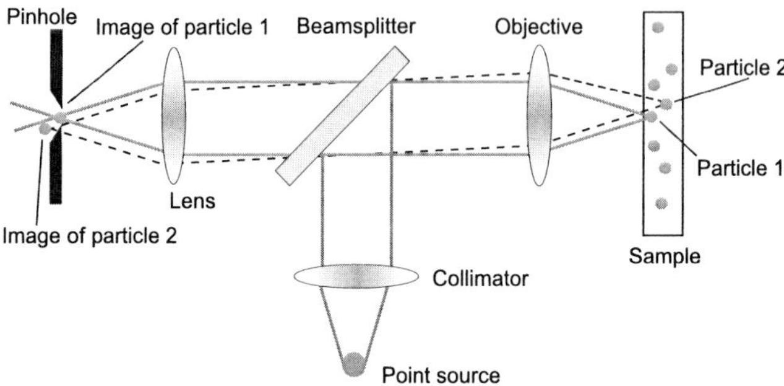

Figure 3.6: *The principle of confocal microscopy.*

In order to get a $2D$ image the sample or the beam needs to be scanned, while a detector behind the pinhole detects the light passing through the pinhole. The advantage of this technique over conventional wide field microscopy is that only a small volume is illuminated and only light coming out of this volume is subsequently detected. In figure 3.6, for example, particle 1 (which could be a scatterer or a fluorescent particle) is imaged on the pinhole, while for particle 2 not only the illumination intensity is much lower, but also its image will not pass the pinhole. The consequence of this imaging technique is the exceptional signal-to-background ratio of confocal microscopes. In particular, it allows to make axial cuts through the sample, a technique which is of

special importance for biological applications. Since such microscopes use point detectors, it is also possible to study extremely fast processes down to the picosecond timescale [ZEK02]. Moreover, the confocal microscope has, as will be shown below, a slightly higher resolution than a conventional one.
The resolution of an optical system is defined via the point-spread function (psf), which is basically the diffraction pattern that arises when a point is imaged through an optical system (lenses, apertures, etc.). The psf of a circular aperture, which is the most common case, has in the focal plane ($\zeta = 0$) approximately the mathematical form of the Airy disc [Hec01]

$$p(\zeta = 0, \rho) = 2j_1^2(\rho)/\rho^2, \tag{3.4}$$

where ζ is the scaled coordinate along the optical axis and $j_1(\rho)$ is the Bessel function of the first kind. It is assumed that all apertures are rotationally symmetric and therefore all directions perpendicular to the optical axis are represented by a second scaled unit ρ. ζ and ρ are defined by

$$\zeta = \frac{2\pi}{N\lambda} NA^2 z \tag{3.5}$$

$$\rho = \frac{2\pi}{\lambda} NA\, r, \tag{3.6}$$

where z is the coordinate along the optical axis and r the coordinate perpendicular to it. λ is the wavelength of the light and $NA = N \sin\vartheta$ is the *numerical aperture* of the objective, where ϑ is defined as the half angle of the cone of light converging to an illuminated spot or diverging from one.
The resolution can thus be defined via the Airy disc by the distance Δx from the central peak to the first dark fringe (Rayleigh criterium)

$$\Delta x = 0.61\lambda/NA. \tag{3.7}$$

According to this criterium, one defines two objects as just resolved if the center of the Airy disc of the first object falls into the first dark fringe of the other. The Rayleigh criterium holds for conventional microscopes as well as for any other optical system.
In the case of a confocal microscope the psf is slightly different. The origin lays in the illumination method. The confocal volume defines the resolution of a confocal microscope. Illumination and detection volumes are described by the same psf, so the volume both illuminated and observed is simply the product of two functions $p(\zeta, \rho)$ [Web96]

$$p_{conf}(\zeta, \rho) = p(\zeta, \rho) \times p(\zeta, \rho). \tag{3.8}$$

The resolution derived from equation (3.8) differs from equation (3.7) because $p_{conf}(\zeta, \rho)$ is a sharper peaked function than $p(\zeta, \rho)$. In the Rayleigh criterium the depth of the dip between just resolved adjacent peaks is 26 %. The same condition leads, for a confocal microscope, to a resolution

$$\Delta x_{conf} = 0.44\lambda/NA. \tag{3.9}$$

The contrast enhancement that discriminates against nearby scatterers in confocal microscopy becomes even more apparent when the obscuring objects are out of the focal plane. The reason can be found again in the properties of the psf. An axial resolution $\Delta\zeta$ can be defined again by Rayleigh's 26 % dip [Web96]:

$$\Delta\zeta = 1.5N\lambda/NA^2. \tag{3.10}$$

For high resolution objectives with an *NA* around 1, the axial resolution is on the order of a wavelength.
The definition of a resolution given here follows the suggestion of Rayleigh; however, this definition is rather arbitrary, always depending on the criteria used. One can easily find several other definitions for the resolution, e.g. the Sparrow-criterium [Hec01] or the definition of a confocal resolution given in reference [CK96]. For evaluation of images with single scatterers or emitters much smaller than the wavelength, which are well separated, the FWHM of the central peak of the psf is a commonly used criteria for the estimation of the resolution (single point resolution) [SHvdV98]. Since the Rayleigh criterium relates the distance between two points, it is a measure for a so-called two point-resolution. Whatever criterium is used, in the end the values are approximately within $\lambda/3$ and λ depending on the *NA* of the objective.

3.2.2 Image generation

The generation of an image in confocal microscopy is carried out by moving the focused spot across the object. As a matter of principle there are two different ways, each method having its own advantages [Web96]. The simpler one is to leave the optics fixed and to move the object with a piezo stage scanner[2]. The other method scans the beam across the sample. As the object does not move, this method is attractive for excitation of nanoparticles on a microsphere, because the sphere is, due to the complexity of the spectroscopy unit (see section 4.2), not moveable.
The principle of beam scanning is to find a position along the optical path in the microscope where a change in angle will result in a linear motion of the focused spot in the object plane. It turns out that for every beam passing the back focal plane (BFP) of the microscope objective under an certain angle, there is a characteristic position of the light spot in the object plane [Paw95]. Since the scanning is usually carried out by mirrors mounted on galvo drives and the BFP is normally located at the entrance pupil of the objective, it is not possible to mount the scanner at this place. A telecentric lens system solves the problem by imaging the pivot point (center of rotation) from a conjugate plane of the BFP into the BFP. In the special configuration shown in figure 3.7, the conjugate plane is placed at the midpoint between two closely spaced scan mirrors. The closer the mirrors are together, the more this arrangement approximates perfect telecentricity. The advantage of this configuration is particularly the small size,

[2]The main drawback of this method is its slowness, rendering it unusable for many modern biological applications.

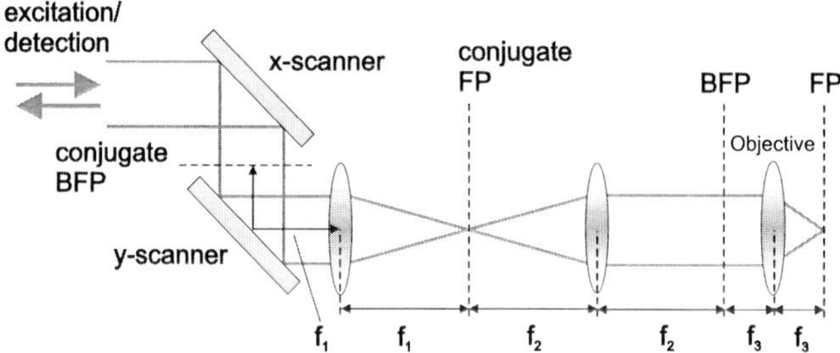

Figure 3.7: *The two scanners change the angle of the incoming light beam in a conjugated BFP and hence change the position of the focus on the sample. The collected light from the sample is de-scanned by the galvos and sent to the detector.*

while the error for scan angles which are not too large is completely negligible. Other possibilities of beam scanning are described in reference [Paw95].

3.2.3 Technical realization

The design of the microscope was subject to several conditions given by the microsphere spectroscopy unit. The microscope needed to operate in a horizontal arrangement with a beam height of $8.5\,cm$ above the optical table, moveable with μm precision in all three spatial dimensions. In order to check the quality, density, etc. of samples it was also necessary to have the possibility of moving the microscope away from the sphere, having enough space to investigate samples on cover slips. The entire microscope is therefore mounted on a composite material baseplate, which can be moved by about $20\,cm$ on an optical rail, while for fine positioning three translation stages can be used. The composite material was chosen, because it saves weight and has at the same time superior mechanical properties. Additionally, some counterweights were added to balance the microscope and to increase the weight for being less sensitive against vibrations (see figure 3.8).
Most of the optical elements are mounted on a small rail, which defines the optical axis. The excitation source is a frequency-doubled Nd:Vanadate laser (Coherent Verdi $10\,W$, $\lambda = 532\,nm$), which is coupled via a single mode optical fiber ($NA = 0.13$) into the microscope. The single mode fiber acts as the point source for illumination, providing a Gaussian beam which is collimated with a lens ($f = 15\,mm$) to a beam diameter of $3.9\,mm$. This beam passes a laser line filter (LLF) just in front of the dichroic mirror (DCM), ensuring that only $532\,nm$ light enters the common illumination/detection path. For scanning the beam, there is a telecentric lens system with $f_1 = f_2 = 100\,mm$, which transmits the beam deflection of the galvos (MediaLas, CT6800HP)

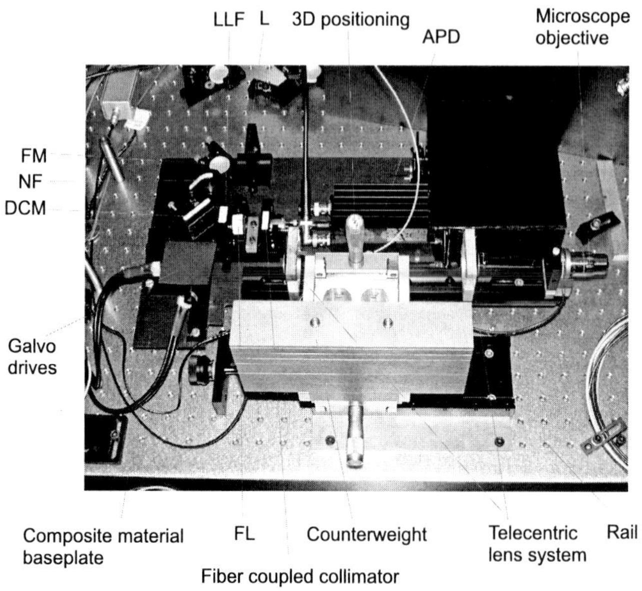

Figure 3.8: *Picture of the beam scanning confocal microscope. Details are described in the text.*

to the BFP of the microscope objective (Olympus ULWD MSPlan 80x, $NA = 0.75$, working distance $4.1\,mm$), equipped with a piezoelectric element for focussing. The beam overilluminates the entrance pupil by 20 %. For focusing down the laser beam to a diffraction limited spot a plane wavefront is needed. Therefore it is common to expand the beam such that its diameter is twice the diameter of the entrance pupil. Nevertheless, it was decided to use this smaller value, in order to keep the setup as simple as possible, especially because ultimate resolution is not the most important characteristic of the microscope. Light from the sample is collected via the same microscope objective, de-scanned by the galvo drives and, in the case of a fluorescent sample, discriminated from the excitation light by an insertable notch filter (NF). The conventional technique with a detection pinhole is circumvented by using a $f = 500\,mm$ lens (L), which focusses the light directly on the avalanche photodiode (APD) (see figure 3.9). The focal length of the lens is chosen such that the focus diameter has the same size as the active area of the APD ($180\,\mu m$). Thus, the APD itself acts as a pinhole. A holder is mounted directly on the APD for additional longpass or bandpass filters, allowing only desired wavelengths to pass to the APD. For further background suppression there is a system of spatial filters constructed as can be seen as a black box in front of the APD in figure 3.8. The light collected by the microscope

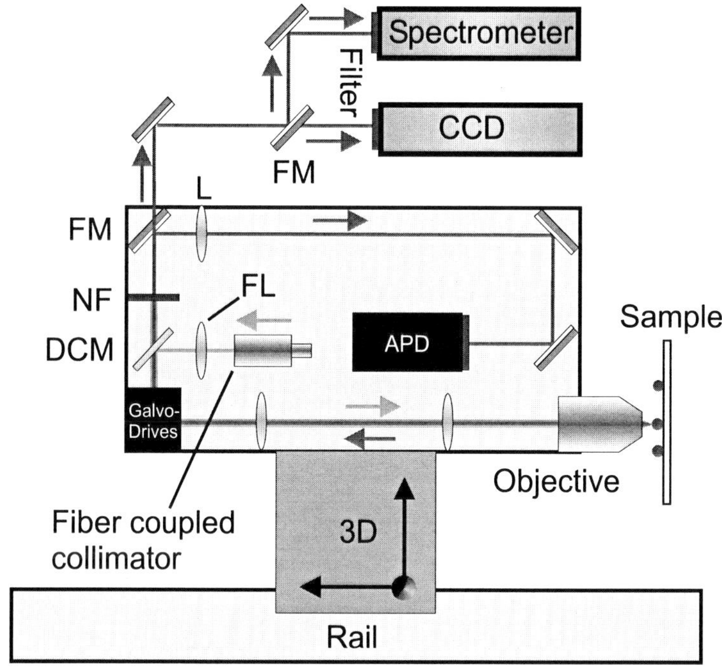

Figure 3.9: *Schematics of the beam scanning confocal microscope.*

objective can also be sent by flipping down a mirror (FM) to an ultra sensitive CCD camera (Hamamatsu ORCA ER), capable to detect the fluorescence of single molecules, or to a spectrometer (Acton Research, Spectra Pro 500i) equipped with a nitrogen-cooled CCD. When using the spectrometer, it is also possible to perform confocal measurements, since the entrance slit together with a software assisted pixel selection procedure forms the detection pinhole. The microscope is designed to be as flexible as possible. It can not only operate in the confocal mode, but one can also switch to a wide field imaging mode, in order to have a real time $2D$ image of the sample, making the selection of a particular area of the sample easy. The switching is done by a single lens ($f = 100\,mm$) on a flippable holder (FL) right after the collimator. This lens focuses the beam in a conjugated BFP and thus into the BFP, leading to a parallel beam illuminating the sample.

In summary, the microscope allows to perform time resolved measurements as well as spectroscopic measurements, both in confocal and wide field mode.

3.2.4 Performance

The microscope was first tested and gauged with a test grating (grating constant $10\,\mu m$) normally used for electron microscopes (see figure 3.10). The field of view in the wide

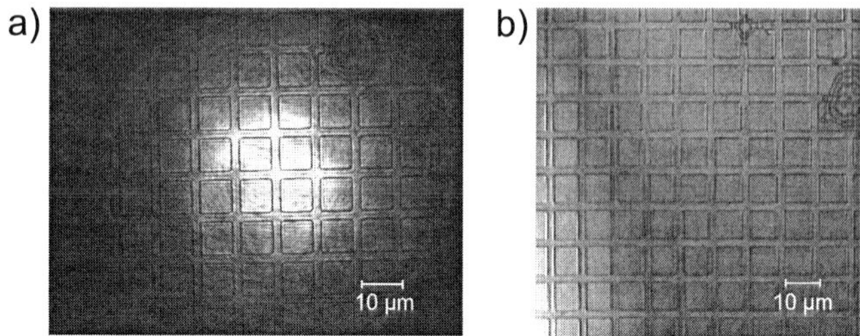

Figure 3.10: *Image of the test grating recorded in a) wide field mode and b) confocal mode.*

field mode is about $(90 \times 90)\,\mu m$, whereas in the confocal mode more than $(350 \times 350)\,\mu m$ can be scanned to get an overview of the sample. However, in figure 3.10 b) one can already see some distortion in the scan, which was carried out a bit away from the optical axis of the microscope. The upper right corner is in this image closest to the optical axis. There are no aberrations evident, leading to a confident scan range of about $150 \times 150\,\mu m$.

In order to further test the performance of the confocal microscope, experiments with CdS/ZnS nanocrystals[3] were performed. These particles were chosen for two reasons. First, it was shown that these particles emit only one photon at a time, thus behaving like quantum emitters, similar to single molecules or atoms [MIM+00]. The detection of such emitters proves the sensitivity of the microscope. Second, due to their small size of about $3\,nm$, they can be treated as ideal point sources, suitable for determination of the resolution of the microscope. Figures 3.11 a) and b) show images of nanocrystals spincoated on a coverslip, the first recorded in wide field, the second in the confocal mode. The integration time of the CCD camera was $150\,ms$, while the time per pixel during the confocal scan was set to be $10\,ms$, resulting in an image acquisition time of $50\,s$ for a $(10 \times 10)\,\mu m$ square. In both cases some individual bright spots are identified, which can be attributed to the photon emission of individual nanocrystals. The resolution was determined in figure 3.11 c) and d) by the FWHM as already described in section 3.2.1 for a single point. The measured values of about $510\,nm$ for wide field imaging and $460\,nm$ for confocal imaging are slightly too big compared with the expected width of $417\,nm$ and $295\,nm$, respectively. These values were obtained by inserting the width of the psf for both modi into equation (3.6). The difference of the

[3]The nanocrystals were provided by the group of Prof. Dr. H. Weller from the University of Hamburg.

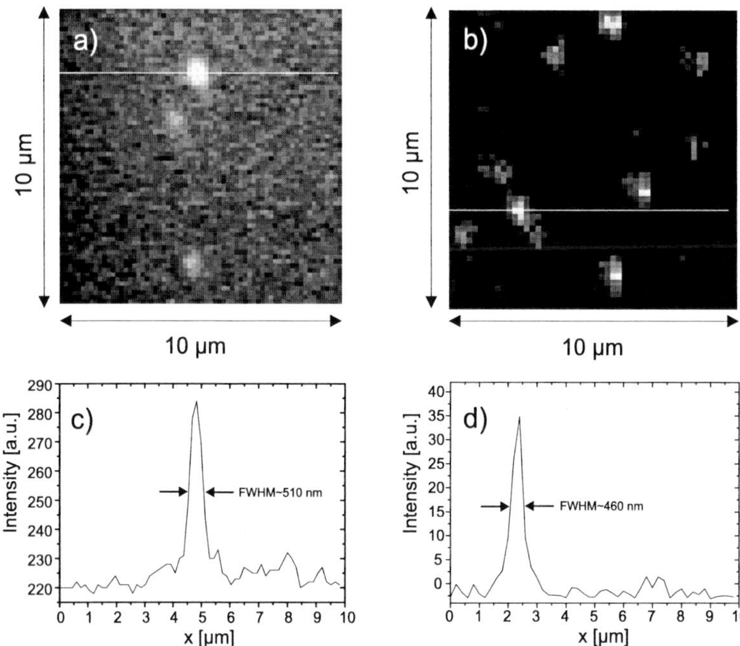

Figure 3.11: *Fluorescence image of nanocrystals on a coverslip in a) the wide field and b) in the confocal modus. c) and d) show cross sections through a) and b), respectively.*

values can be explained by the deviation of the real microscope from ideality, where the objective with a certain NA is considered to be the limiting element. However, in the real system other factors such as imperfect collimation, insufficient over illumination of the entrance pupil of the objective and lens errors can decrease the final resolution. Using achromats for collimation and for the telecentric optic should already improve the performance. But for the special application in this work, where the microscope was used as a flexible tool for addressing a single nanoparticle on a microsphere, the achieved spot size was sufficient.

Two experiments confirm that figure 3.11 indeed shows single crystals and not agglomerates of several nanocrystals. The emission properties of one of these fluorescent spots were investigated in the time and in the spectral domain. The confocal microscope allows to position the focus of the laser beam (excitation power density $150\,W/cm^2$) on one and only one of them. Figure 3.12 a) shows the temporal behavior of the emitter, measured with the APD. One can clearly see the characteristic "on" and "off" behavior, well known as "blinking", which is typical for single nanocrystals [NDB+96, SBPM02]. During the "on" time the intensity has not necessarily a constant value, as one might first expect. The reason for this is that one always integrates over a certain time[4], given by the resolution of the detector. Since nanocrystals blink on any timescale [MHG+01], one can expect to have different numbers of photons in each integration interval. Additionally, the spectral characteristics of the emitter strongly

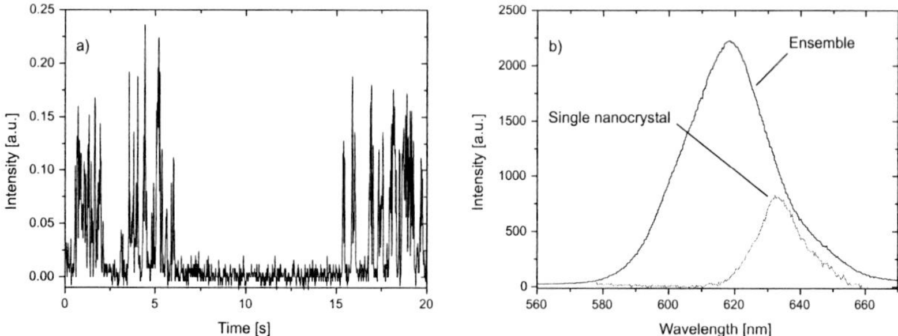

Figure 3.12: *a) shows the well known "blinking" (fluorescence intermittency) behavior of a single nanocrystal. In b), the spectrum of a single nanocrystal is plotted under the envelope of an ensemble spectrum. The width is about half the width of the ensemble.*

indicate that one deals with a single nanocrystal. Figure 3.12 b) shows the spectrum of an ensemble of nanocrystals and what is believed to be that of a single one. The spectral width of the ensemble is $27\,nm$ centered around $618\,nm$, while the width of the single nanocrystal was determined to be only $14\,nm$, with the center around $633\,nm$. Further measurements on other nanocrystals revealed the well known variation concerning the center peak of single nanocrystal spectra within the envelope of the ensemble [ENSB99, NSW+00, SBPM02]. Notice that the spectrum of the single nanocrystal in figure 3.12 b) is still inhomogeneously broadened, due to the spectral diffusion within the integration time of $30\,s$. This phenomenon is widely discussed in literature [NSW+00]. Linewidth measurements and blinking are nowadays well accepted indicators for identification of single nanocrystals, but strictly speaking an ultimate proof can only be given by the measurement of the second-order correlation function $g^{(2)}(\tau)$ (Hanburry Brown-Twiss experiment), as it was done in references [MIM+00] and [LBG+00].

[4]The time resolution in the experiment was never less than $1\,ms$ due to the electronics behind the APD.

3.3 The scanning near-field optical microscope

In the last section, the basic concepts of ultra sensitive high resolution microscopy were introduced. The techniques described there were based on far field imaging, where the resolution is principally restricted to about $\lambda/2$ due to diffraction. However, this fundamental limit could be overcome, if one accesses the optical near-field. Although this idea was already proposed in 1928 [Syn28], the first experimental realization of a **S**canning **N**ear-field **O**ptical **M**icroscope (SNOM)[5] was only in 1984 [PDL84, LIHM84]. SNOM is a scanning probe technique similar to Atomic Force Microscopy (AFM). One obtains optical images with resolutions well beyond the diffraction limit, typically between 50 and 100 nm. Some publications even report values of 20 nm [DPR86] and 12 nm [BTH$^+$91], respectively.
For this work, the SNOM turned out to be a powerful tool for mapping the evanescent fields outside the microsphere for mode identification. But the main reason for the SNOM's construction was its use as a nanotool to position a single nanoparticle with nm accuracy into the evanescent field of a high-Q whispering-gallery mode.

3.3.1 Basics of near-field imaging

A thorough and quantitative treatment of near-field imaging is still a challenging task. As pointed out in reference [San01], the solution of the Maxwell equations with rough surfaces as boundary conditions is a very demanding job, especially because analytical or semi-analytical approaches cease to be valid under certain circumstances and numerical calculations are much too heavy for a Pentium cluster. Therefore, a rigorous treatment cannot be given here. The interested reader is referred to review articles by Greffet and Carminati [GC97] and by Girard and Dereux [GD96]. Here only a simple approach using Fourier optics and some aspects for experimental work can be given.
The main problem of imaging theory is how an **E**-field distribution $\mathbf{E}(x, y, z_0)$ in the object plane (at $z_0 = 0$) propagates to the image plane at $z = z_i$. Now, any distribution $\mathbf{E}(x, y, z_0)$ can be written in terms of Fourier components in the xy-plane $e^{i\mathbf{k}_{\|}\mathbf{r}_{\|}}$, where $\mathbf{r}_{\|} = (x, y, z_0)$ and $\mathbf{k}_{\|}$ denotes the different spatial frequencies parallel to the surface. As in the case of a usual Fourier decomposition, it is necessary to have high spatial frequencies to be able to express small structures. For a monochromatic wave with $\lambda = \frac{2\pi}{k_0}$ the following expressions hold:

$$k_0^2 = \left(\frac{2\pi}{\lambda}\right)^2 = \frac{\omega^2}{c^2} = \mathbf{k}_{\|}^2 + k_z^2 \tag{3.11}$$

or

$$k_z = \sqrt{\frac{\omega^2}{c^2} - \mathbf{k}_{\|}^2} \tag{3.12}$$

where ω is the optical frequency and c speed of light. k_z denotes the component of the wave vector perpendicular to the surface. k_z becomes pure imaginary, if $|\mathbf{k}_{\|}| > \omega/c$.

[5] The abbreviation SNOM is usually also used for **S**canning **N**ear-field **O**ptical **M**icroscopy.

This means (compare section 2.4) that the field is evanescent and therefore not propagating in the z-direction towards the image plane. Therefore, any information on small structures that require $|\mathbf{k}_{\parallel}|$ larger than ω/c is not carried into the far field and thus do not contribute to the image generation. This condition shows the origin of the resolution limit of about $\lambda/2$ in conventional microscopy. Near-field techniques, however, can use these evanescent modes for imaging. The relevant mechanism for high resolution scanning near-field optical microscopy is scattering. The evanescent modes need to be converted into propagating modes. This is usually done by placing a sharp optical fiber tip, which acts as the near-field probe, close to the surface of the sample. Different SNOM configurations using fiber tips are sketched in figure 3.13 [Oht98, Kra02].
The distance between tip and sample surface acts on the electrical field like a lowpass, giving a condition (rule of thumb) for the resolution of SNOM: an object can be detected with a resolution d, if the tip-surface distance is smaller than $d/2\pi$, which defines the extension of the near-field. A second criterium that has to be considered when talking about the resolution of a SNOM is the tip geometry, especially the end diameter of the tip and the aperture size in aperture SNOM, respectively. Both end diameter and aperture size determine, again as a rule of thumb, the resolution of the SNOM. The discussion above also suggests a definition of the "near-field", which was not yet given. In the context of superresolution, the near-field can be defined as the region where the evanescent waves contribute significantly to the field [GC97].

3.3.2 SNOM configurations

It was already mentioned above that the essence of SNOM is the local scattering of the evanescent field. Several modes with different illumination and detection methods are nowadays established, each having its own advantage depending on the application. The most commonly used methods are sketched in figure 3.13 [Oht98, San01].
In aperture SNOM a tapered optical fiber is aluminum coated such that at the very end a very small subwavelength opening is left [BT92]. This aperture realizes a nanoscopic light source which illuminates the sample (see figure 3.13 a)). At the sample the evanescent field created by the aperture is scattered into the far field and then detected. It is also possible to use the same fiber for excitation and collection if fluorescent samples are investigated [SM99].
In another configuration (scattering SNOM) the sample is illuminated globally and the evanescent fields are scattered by a fiber tip [ZMW95] (see figure 3.13 b)) or a metal coated AFM cantilever [KK97] into the far field. Since the near-field probe itself scatters the excitation light strongly, usually modulation and lock-in techniques need to be applied to discriminate the near-field signal from the background.
In the so-called photon tunneling SNOM [CBB94] a transparent sample is illuminated via total internal reflection as shown in figure 3.13 c). The advantage of this technique is in its quiet background. Only the signal that is scattered at the junction between the tip and the sample enters the fiber and reaches the detector. This mode was used

in this work to probe the evanescent field at the surface of the microsphere and to develop a scheme for mode identification (see section 4.4).
Aperture SNOM is not the only possibility for realizing a nanoscopic light source. A fluorescent nanoparticle (e.g. a diamond color center [KHS+01] or a single molecule [MHMS00]) perfectly realizes such a light source and can thus be used for sample illumination (see figure 3.13 d)). The nanoparticle can be excited in the far field. Here a peculiar feature is that only red shifted fluorescence contributes to the image formation, resulting in a high signal-to-noise ratio. The fluorescent emitter is preferentially attached to the end of a near-field probe which allows a precise scanning.

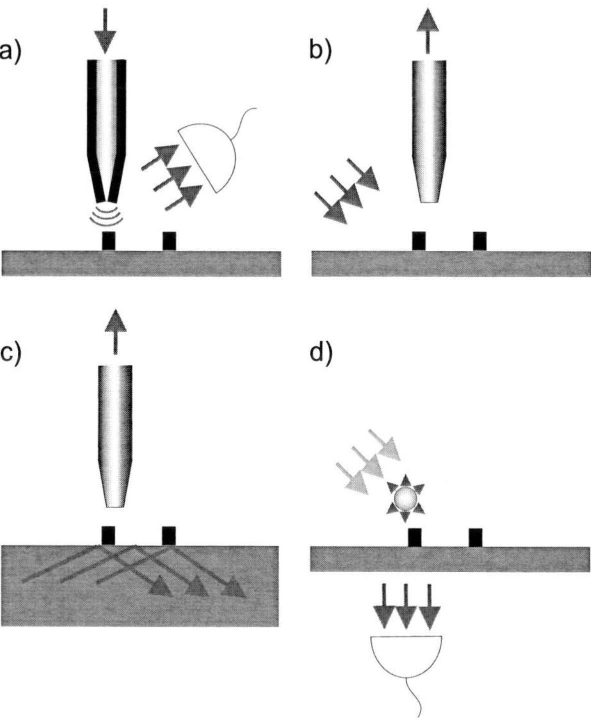

Figure 3.13: *Different SNOM modes: a) aperture SNOM, b) scattering SNOM, c) photon tunneling SNOM, d) active probe SNOM. The last three methods are different realizations of so-called apertureless SNOM.*

3.3.3 Shear-force mechanism for distance control

As pointed out above, the distance between the fiber tip and the sample's surface is a crucial parameter for the resolution of a SNOM. Therefore, a method is needed which provides a distance control on the nanometer scale between the fiber tip and the sample surface during the scan. Measuring the tunnel current between the sample and a metallized tip is one method [LL93], but it is restricted to conducting samples.
A wide spread method is another technique, the so-called shear-force feedback control [BFW92], which is a distance control based on force microscopy. A fiber tip is excited to lateral oscillations of a few nanometers, while the amplitudes of these oscillations are detected at the same time. Approaching the tip to a surface leads, due to the interaction of the tip with the sample, to a damping of the amplitude and to a phase shift. Both signals can be used to provide a feedback [BFW92], which is usually sent to a piezoelectric element. Keeping, for example, the amplitude of the tip's oscillation constant will thus stabilize the tip-surface distance to a certain value, which is typically about $10\,nm$. Moving the tip around and recording the piezo voltage leads directly to a topography image of the sample. The physical origin of the shear-force mechanism is still controversially discussed. Under ambient conditions, it is believed, that a few nanometer thick water layer, present on every surface, damps the oscillation [BMH99]. However, shear-force works also in the absence of this water layer, under cryogenic conditions as well as in ultra high vacuum chambers. In these cases a totally different mechanism, a mechanical knocking [GBSU96], appears to be responsible for the shear-force damping. The interaction range, defined as the distance at which the oscillation amplitude is damped from $100\,\%$ to $0\,\%$, has an extension between $2\,nm$ and $30\,nm$, depending on the ambient conditions and oscillation amplitude.

3.3.4 Technical realization

There are several methods to implement the shear-force distance control in a SNOM. Here, a non-optical technique was favored due to the compactness of the scheme. A segmented piezo tube is used for both excitation and detection of the oscillation of the fiber tip [BHM96]. The tips were produced either from single-mode optical fibers (cladding diameter $125\,\mu m$) with a commercial pipette puller (Sutter Instruments P-2000) or by an etching process with hydrofluoric acid. Both procedures are well described in references [Oht98, Weg98, Kal02]. The mechanical resonance of the tip is determined by the shape of the tip and its length sticking out of a micropipette. The fiber is fixed inside the micropipette, either with cyanoacrylate glue [Kal02] or by melting the pipette having a low melting point around the optical fiber in a focussed CO_2-laser beam. The tip, sticking out can then be modelled by the cantilever beam.
A length between $2.4\,mm$ and $2.5\,mm$ leads to a resonance frequency around $25\,kHz$ having a Q-factor around 100 in the case of glued tips and up to 1000 for fiber tips prepared with the CO_2-laser procedure. For excitation, the micropipette is put into the

segmented piezo together with some polymer of high viscosity (BASF Oppanol B3), which fills the residual gap between the two (see figure 3.14).

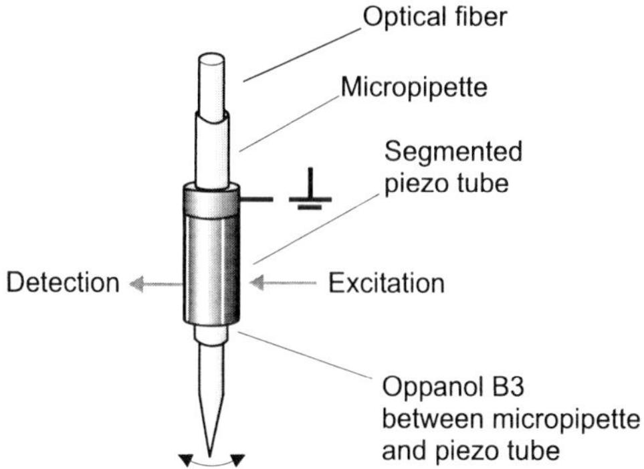

Figure 3.14: *Excitation and detection of the fiber tip oscillation with a segmented piezo tube.*

The polymer acts like a mechanical highpass, getting stiff at frequencies $> 1\,kHz$ and thus transmitting the vibrations of the piezo segment to the fiber tip. For stability reasons the micropipette has to be clamped with a screw behind the piezo tube, as can be seen in figure 3.15.

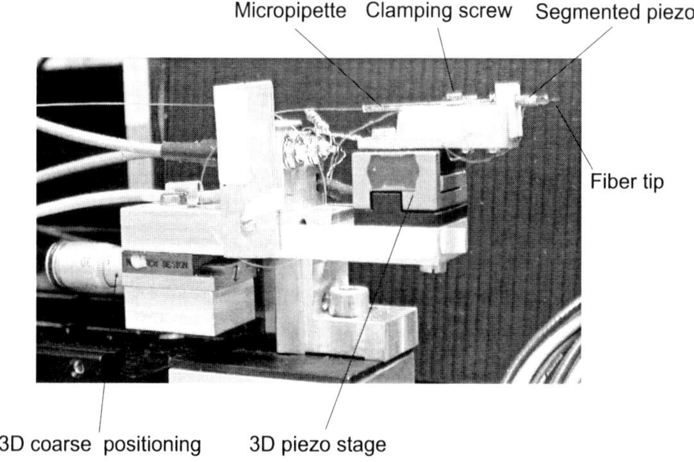

Figure 3.15: *Photograph of the SNOM.*

The excitation of the tip is carried out by applying an AC voltage of $10\,mV$ between the inner grounded electrode and one of the outer segments. When the segment is driven at the tip's resonance frequency, its oscillation amplitude is strongly enhanced. Reference [Kal02] reports as an upper limit for the oscillation amplitude of the tip a value of $13\,nm$ for a Q of 100 in a SNOM very similar to the one used in this work. The deflection of the detection segment is large enough, by this oscillation, to be detected by measuring the induced voltage between this segment and the inner ground electrode. The obtained signal is first amplified and filtered with a low noise preamplifier and then demodulated with a lock-in amplifier where the function generator which is used to excite the fiber tip acts as the reference. Figure 3.16 shows the signal detected when the frequency is swept over the resonance.

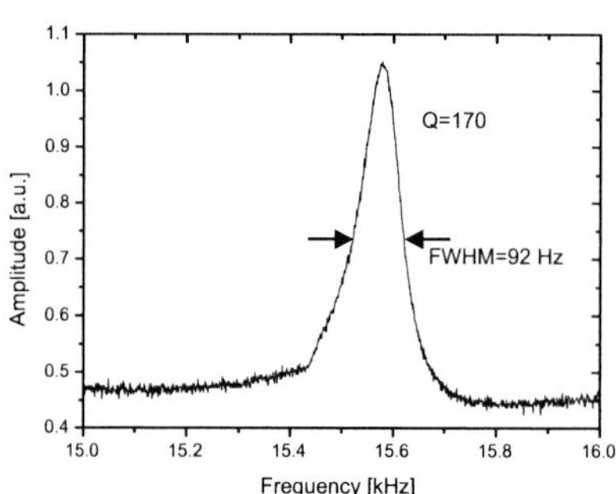

Figure 3.16: *Resonance of a typical near-field probe.*

To obtain an image, the tip has to be scanned, while a photomultiplier tube (Hamamatsu 5784−04) detects the light collected with the fiber tip. During the scan the surface-tip distance needs to be held constant, which is done with a PI-controller, P representing the proportional and I the integral gain. It delivers a control output, which is first amplified by a high-voltage amplifier and then sent to the actuator (z axis of the 3D piezo stage in figure 3.15), regulating the tip-surface distance (see figure 3.17). Between servo and amplifier was an electronic notch filter, combined with a lead phase correction, inserted in order to suppress gain at the piezo resonance and to lift up the phase at lower frequencies, for better performance of the control loop[6]. According to section 3.3.3 the amplitude of the resonance acts as the control variable in the feedback loop. The working point is set slightly below the resonance frequency. At a value of about 90 % of the maximum of the resonance the sensitivity is highest, because the resonance gets broader as it is damped when the tip approaches the surface, like in the case of a damped harmonic oscillator. Additionally, the resonance frequency is shifted to a higher value, a fact which is explained by an increase of the spring constant of the tip in the shear-force interaction zone. Both effects make sure that the working point is now in the edge of the resonance, where the sensitivity is maximal.

In order to calibrate the piezo element axes and to test the performance, a test structure with a grating constant of $3\,\mu m$ and a step height of $485\,nm$ was scanned with

[6] An introduction into the properties of servo loops can be found in the book by Tietze and Schenk [TS02].

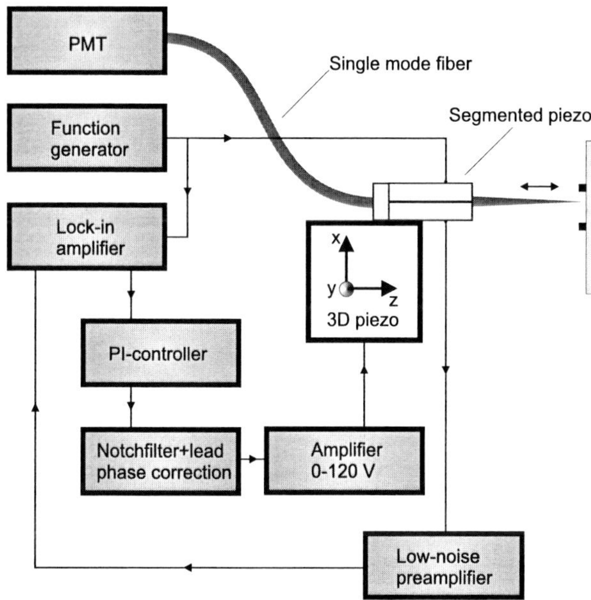

Figure 3.17: *Block diagram of the shear-force control loop. The optical detection is also sketched: the fiber tip is directly coupled to a photomultiplier tube (PMT).*

a speed of $2\,\mu m$ per second (see figure 3.18). Although such structure heights are already challenging for a SNOM scanned with such a speed, the tip follows the structure without major difficulties, mapping the topography of the sample. Since the structure is rather deep, the sidewalls of the steps appear not to be perfectly vertical. The reason for this is simply the conical shape of the fiber tip. The upper edge of the step interacts with the side of the tip, forcing the feedback loop to pull the tip backwards and thus displaying the fiber tip's shape instead of the step's exact shape. This simple example demonstrates how artifacts enter shear force images. To get an idea about the topography resolution, the standard deviation of the mean value of every plateau in figure 3.18 was taken, resulting in a value of less than $3\,nm$. This is a pessimistic estimate, since the plateaus are not perfectly horizontal (linear background already subtracted). The recorded data points have therefore a systematic deviation from the mean value. The lateral resolution is given by the tip's end diameter, which is on the

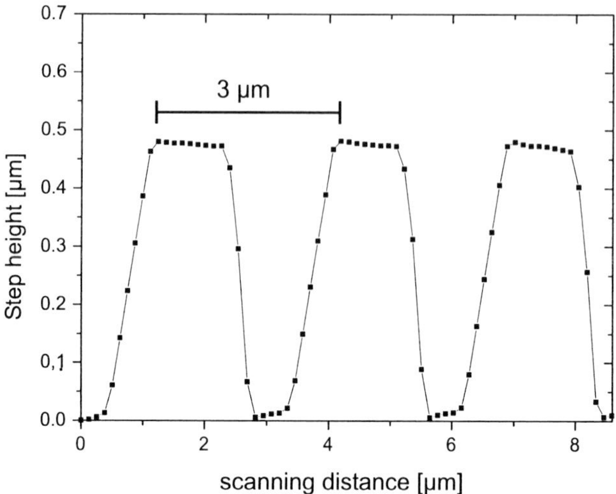

Figure 3.18: *Gauging the SNOM: cross section of a test sample with a periodicity of* $3\,\mu m$ *and a step height of* $485\,nm$.

order of $100\,nm$[7]. To give a number for the optical resolution of the SNOM is difficult, because of the dependency of the resolution on the fiber tip's exact shape. However, the standing wave pattern formed by the whispering-gallery modes in the microspheres could be used later in this work to give an upper limit for the resolution (see section 4.4), which was better than $210\,nm$ for the particular fiber tip used in the experiment.

[7]The end diameter of the fiber tips was determined with a Scanning Electron Microscope.

Chapter 4

Microsphere characterization

This chapter describes the fabrication of silica microspheres and the experimental setup for spectroscopy of the whispering-gallery modes present in these spheres. This spectroscopy unit is the heart of the experiment and is essential for all more sophisticated experiments in the following chapters. It allows not only the determination of the Q-factor and the modes identification, but the setup also allows the tuning of the Q-factor over three orders of magnitude and the variation of the photon outcoupling efficiency out of the sphere by adjusting the tunneling gap between the sphere and the prism coupler. The SNOM, discussed in the previous chapter, is used for unequivocal identification of the fundamental mode and to optimize the coupling condition of this mode to the prism coupler in a unique way. Furthermore it is possible to characterize the sphere's geometry, for example its ellipticity.

4.1 Fabrication of silica microspheres

For the fabrication of silica microspheres with a very high Q-factor, it is absolutely necessary to start with extremely clean feedstock. Any kind of contamination will degrade the Q. In the ultraviolet metallic impurities reduce the Q, while in the infrared water contamination can spoil the Q substantially. High quality spheres were produced from $3\,mm$ diameter Suprasil 300 glass rods (Heraeus Quarzglas GmbH) with an extremely low ion contamination (e.g. $OH^- \leq 1\,ppm$). These rods were cleaned in a three step procedure. In the first step the rods were put in a solution of KOH-isopropanol for about $30\,min$. This solution dissolves almost any kind of organic material from the exterior of the rods. Then the rods were put for another $30\,min$ in nitric acid $40\,\%$, which is known to be a strong acid and a powerful oxidizing agent, in order to remove residual chemicals from the silica surface. These two steps assure that the rods are clean, such that in the last step a $40\,\%$ hydrofluoric acid solution can etch an entire layer of silica away, since water diffusion can contaminate the rods over months of storage [Str99]. The thickness of the removed layer depends strongly on the age and the initial dirt on the surface of the rods. By etching almost $1\,mm$ away, the results were always satisfactory. The clean rods were then pulled in the flame of an oxy-hydrogen torch to thin fibers and cut into small pieces, which were glued with a cyanolit glue into small copper pipes (inner diameter $200\,\mu m$, outer diameter $1\,mm$, length $23\,mm$). These copper rods could be fixed on a 3D translation stage to position the fiber in the focus of a CO_2 laser beam (see figure 4.1). The fibers easily melt under irradiation of the

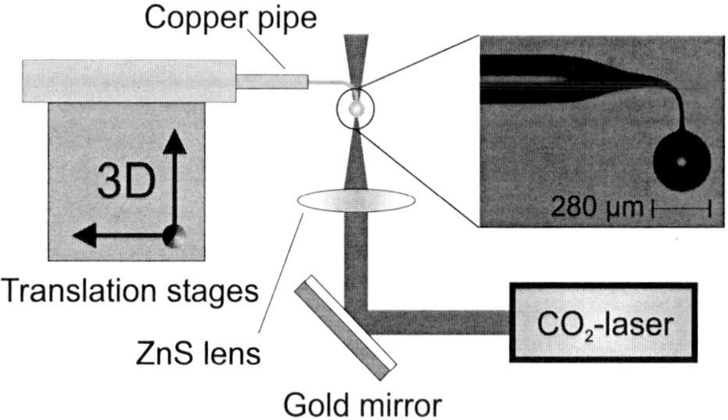

Figure 4.1: *Sketch of the microsphere fabrication setup. The picture shows a $280\,\mu m$ sphere.*

CO_2 laser (Synrad G 48-1, $P = 10\,W$), because its emission wavelength of $\lambda = 10.6\,\mu m$ is strongly absorbed by the glass. First, a rod is pulled to a desired diameter using tweezers, then gravity was used to bend it into a hook for later manipulation of the microsphere in the experiment. Surface tension finally forms the sphere at the end of the fiber, when it is melted in the focus of the laser beam. With this procedure silica microspheres were routinely produced, having diameters between $40 - 300\,\mu m$, adjustable during the melting process. The entire procedure was overseen with a long distance microscope (Olympus SZ11). Figure 4.1 shows a picture of a large sphere. An atomic force microscope measurement confirmed the quality of the fabrication process, since the measured surface roughness was about $0.5\,nm$, which was the resolution limit of the used device.

As already stated in section 2.3.2, the sphere can not be treated as ideal for high-Q modes. The stem acting as a holder is useful for manipulation but breaks the symmetry of the sphere and lifts the degeneracy of modes with different m numbers. The sphere is therefore slightly elliptic having a defined equatorial plane, approximately perpendicular to the stem. The fundamental mode can only exist in this plane [GI94], therefore a precise adjustment of the sphere with respect of the prism is necessary, if one wants to excite this mode efficiently.

4.2 The microsphere spectroscopy unit

For the excitation of the whispering-gallery modes a setup was constructed, based on a prism as input and output coupler. Light from a tunable, narrow linewidth diode laser passes an optical diode (Linos DLI-2, 60 dB), which protects the diode laser from unwanted back reflections (see figure 4.2). Then the beam enters an **E**lectro-**O**ptical **M**odulator (EOM)(Linos LM0202), driven at a frequency of 11 MHz to create side-bands on the signal to have a gauge, in order to measure high Q-factors precisely. Since microspheres easily show optical bistability as a thermal effect when light is efficiently coupled [BGI89], a $\lambda/2$-plate in combination with a polarizing beam splitter (PBS) is used to adjust the intensity to a level at which this effect is not present anymore. A single mode optical fiber acting as a spatial filter finally delivers the laser radiation to the experiment. The linear polarization is kept with a fiber polarization controller, while the rotation of a $\lambda/2$-plate in front of the fiber coupler allows a controlled turn of the polarization axis and thus a selective excitation of TE and TM modes, respectively. The spectroscopy unit itself is covered by a perspex box protecting the sphere against dust particles. Furthermore, it helps to keep the temperature of the sphere stable. The laser beam, which exits the fiber as a Gaussian beam, is collimated and then focussed under a certain angle on the prism (see section 2.4). Since this angle is bigger than that of total internal reflection, the beam gets reflected. A lens collimates the beam before it leaves the box and hits a photodiode. During the experiment the diode laser needs to be monitored. Right after the optical diode some light is branched off the main beam with a beam splitter (BS) and divided between a wavemeter (High Finesse Ångstrom-WS/6, resolution 600 MHz @ 670 nm) and a home built confocal Fabry-Perot ($\mathcal{F} = 150$, modespacing 3 GHz). The Fabry-Perot is scanned $30 - 50$ times faster than the diode laser in order to monitor the modehop free operation of the laser. Furthermore, it is used to gauge the computer software, which scans the laser, while recording the signal from the photodiode and the fiber coupled photomultiplier tube (PMT). The wavemeter allows first to determine the frequency of the whispering-gallery mode, so that it can easily be found again at a later time and secondly, it is used to put several scans together in order to span a wider range in frequency space. This function is especially useful, if one wants to look systematically for a certain mode, because then one needs to span up to a FSR without any gap.

The spectroscopy unit in the perspex box is shown in a simplified schematics in figure 4.2. Figure 4.3 gives a photograph, in order to visualize the number of degrees of freedom needed to optimize the coupling to high-Q whispering-gallery modes. The single mode fiber is mounted with the collimator and the focussing lens (Achromat, $f = 10\,mm$) as a compact monolithic unit to a socket. This socket is actually a small tower with two translational and two rotational degrees of freedom. It allows a precise adjustment of the optimal launching angle (one rotation of the screws corresponds to a change of one degree), which is, according to section 2.4, a horizontal beam hitting the prism under an angle of $\Psi = 59°$. The fine adjustment is performed with a micro-

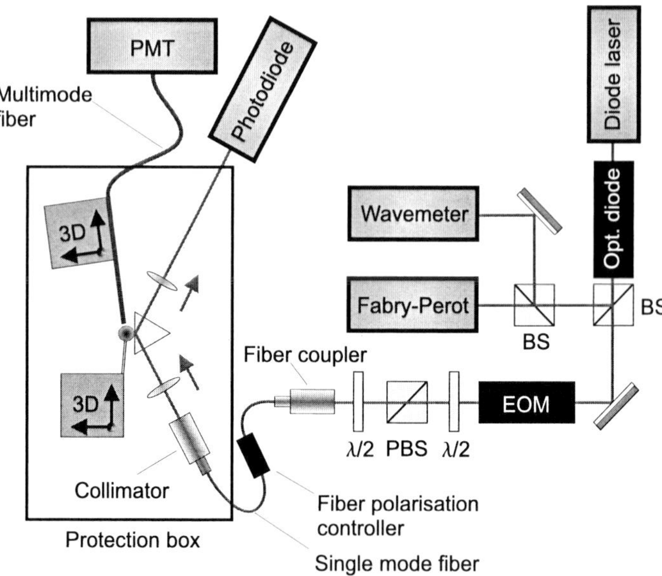

Figure 4.2: *Sketch of the experimental setup for efficient excitation of whispering-gallery modes.*

sphere coupled to the prism, by optimizing the coupling to a high-Q mode [GI94]. The translation stages allow to focus the beam on the prism and to position the focus on a desired position on the prism. The focus is minimized, either by imaging the prism surface with the microscope in the wide field modus (some surface inhomogeneities always scatter some light) or with the SNOM which is perfectly suitable to map the evanescent field created by the focussed laser beam [SBK+03]. The fiber tip is scanned over the focus, while a photomultiplier tube (Hamamatsu $5784-04$) records the signal at the other end of the fiber. The focus diameter, defined as the $1/e$-width was determined to be $4.5\,\mu m$ in the vertical direction and $13.8\,\mu m$ in the horizontal direction. The elliptic shape is to be expected, due to the angle of incidence. The horizontal extension is, however, $65\,\%$ larger than expected, a fact that could be explained by aberrations induced when the beam has an inclination angle in this direction on the lens or passes the lens off axis. This hypothesis was supported by a careful inspection of the alignment of the focussing lens.

For accurate positioning of the microsphere in the evanescent field created by the focussed laser beam, a stage having three translational and two rotational degrees of freedom was constructed. A long distance microscope (Olympus SZ11) was used to observe the microsphere's course positioning with a three axis manual translation

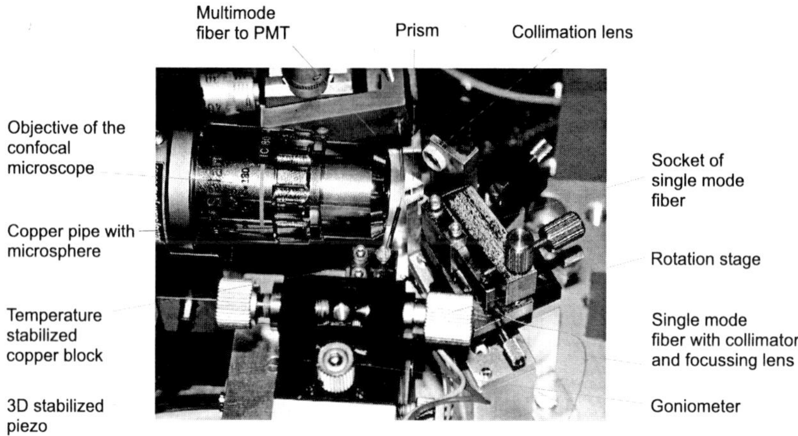

Figure 4.3: *Photograph of the microsphere spectroscopy unit. The size of the prism is* $5\,mm$.

stage having a accuracy of $1\,\mu m$.

The fine positioning is done with a 3D piezo cube having strain gauge (Piezosystem Jena, Tritor 38 SG). The active position control with the strain gauge is necessary for the axis approaching the sphere to the prism, since piezoelectric elements tend to drift and a change in the sphere prism gap of only a few tens of nanometers can change the Q-factor, as well as the in- and outcoupling efficiency substantially. For a controlled and reproducible coupling to the fundamental mode it is of great importance to tilt the sphere in such a way that its equatorial plane is horizontal, as the plane defined by the focussed laser beam is. If this is not the case light does not couple to the fundamental mode, which can be seen from geometrical considerations as given in section 2.3.2. Therefore, a goniometer combined with a rotation stage allows to rotate and to tilt the sphere with a resolution better than one degree.

The copper rod itself is mounted with the sphere in a copper block which is temperature stabilized. This feature was included in the setup, since the sphere changes its resonances frequencies by about $2\,GHz/K$, mainly due to changes in the refractive index [SB91]. To resolve the high-Q modes with a width of a few MHz or smaller well enough, it is necessary to zoom the scan range of the diode laser around a resonance down to $50\,MHz$. A change in temperature of only $15\,mK$ thus shifts the resonance out of the scan range. The temperature stabilization allows also to change the temperature intentionally, as a method to tune the frequency of a certain whispering-gallery mode to a desired value.

4.3 Determination of the Q-factor

Spectroscopy of whispering-gallery modes is the starting point of every experiment with the microspheres. It allows not only to determine the intrinsic Q-factor of the microsphere, but also to optimize the coupling conditions for incoupled and outcoupled light. By varying the gap between sphere and prism, the coupling efficiency and Q-factor can be substantially altered in a controlled way. While the diode laser is continuously scanned, the microsphere is positioned in the evanescent field, created by the laser beam focussed on the internal face of the prism. Whenever the laser is in resonance with the sphere, the total internal reflection gets frustrated and photons tunnel into the sphere, building up a mode. The photodiode detects the resonance as an absorbtion dip. Additionally, the fraction of light in the resonant mode, which is scattered out of the sphere, is collected by the multimode fiber and guided to the PMT (see figure 4.2). Absorption and scattering signals are shown in figure 4.4 for a scan width of $8\,GHz$. The detection with the photomultiplier results in a high signal-to-

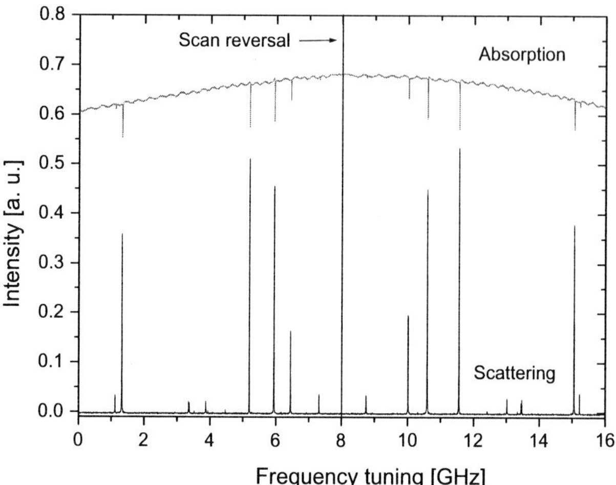

Figure 4.4: *A mode spectrum of a sphere in absorption (upper trace) and scattering (lower trace). The laser frequency was tuned by* $8\,GHz$ *[GDBS01].*

noise measurement of the mode spectrum. The sensitivity is superior to the absorption measurement with the photodiode, even when lock-in techniques were used to increase the sensitivity of the absorption measurement. Weakly coupled modes easily detected with the PMT could not be found in the absorption signal. Therefore, only the scattering signal was usually used to find the whispering-gallery modes and to measure

the Q-factor. The absorption signal was only evaluated when the coupling efficiency needed to be optimized and measured. The highest coupling efficiency achieved was 80 %, as good as the best value reported in literature for a prism coupler [BGI89].
Figure 4.5 a) shows the spectrum of a microsphere recorded over $60\,GHz$. One can clearly see the selectivity of the prism coupler. The sphere is adjusted such that the coupling for the modes with $m \approx l$ is strongest, which means for figure 4.5 a), that resonances with a lower frequency have a smaller difference in $l - m$. Tilting the sphere allows to change the coupling efficiency to the case where higher order modes are stronger excited than the fundamental mode with $l = m$. The first six modes seem to be equally spaced, as one expects for modes with $m \approx l$ belonging to the same family (see equation (2.55) and assume an l of several hundreds). The direct proof of the affiliation of modes to the same family will be given in section 4.4.
For a precise measurement of high Q-factors, it is necessary to zoom into an area around a mode, as is done for the second mode in figure 4.5 a). Two different techniques are used, in one the EOM creates sidebands as a frequency ruler, in the other the diode laser scan range is gauged with the Fabry-Perot. The resonance is first fitted, according to section 2.1, with a Lorentzian lineshape (or double Lorentzian) as in Figure 4.5 b) and c). The obtained FWHM is then put into equation (2.7) to determine the Q-factor, where the frequency of the laser was determined with the wavemeter. The technique using the EOM has an accuracy of about $100\,kHz$, given by the standard deviation of the FWHM. For the second method, it is difficult to tell an absolute accuracy, due to the nonlinear behavior of the piezoelectric elements used to scan the diode laser. Instead a relative accuracy of 50% with respect to the method using the EOM can be given. The Q-factor of the mode determined by the FWHM obtained in Figure 4.5 b) and c) is 3.0×10^8 and 4.5×10^8, respectively. Q-factors $\geq 10^9$ were measured in this work, probably still limited by the linewidth of the diode laser (see section 3.1.3). The mode shown in Figure 4.5 is split, an effect well known for high-Q whispering-gallery modes [WSH$^+$95, GPI00, KSV02]. The origin of that splitting is internal backscattering that couples the two degenerate whispering-gallery modes propagating in opposite directions inside the sphere (modes only differing in the sign of m). Backscattering induced by small surface defects, defects in the silica network or residual impurities having an intensity of 10^{-10} per round trip is enough to explain the observed splitting [WSH$^+$95]. This splitting can vary from 0 to several MHz. It was observed that sometimes only one of two consecutive modes of the same family, both having Q-factors of several times 10^8, showed a splitting. The reason can be found in the different intensity distributions of the two modes (see section 2.3.2). For example, a surface defect located at the maximum of the intensity distribution of the fundamental mode would split that mode, but would not affect the mode with $l - |m| + 1 = 2$.

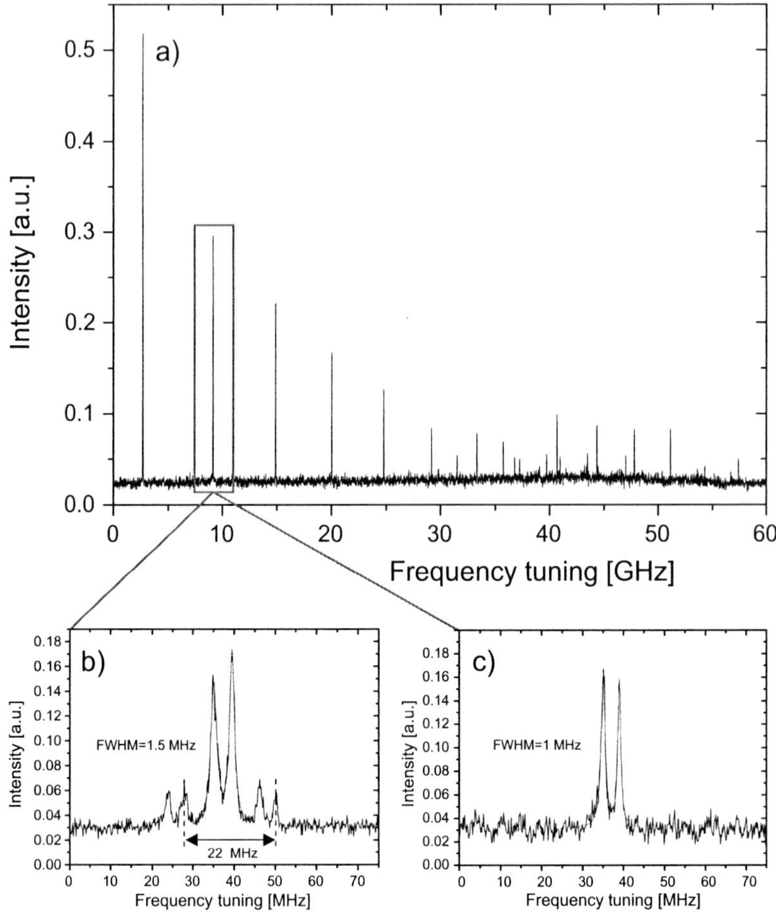

Figure 4.5: *Spectroscopy of a microsphere: figure a) shows a spectrum over* 60 GHz. *The first six modes are assigned to the same family. Then other modes appear, making the assessment more difficult. One can also see the selectivity of the prism coupler. Figures b) and c) show a zoom around the second resonance for a Q-factor measurement. In b) an absolute frequency marker is generated by sideband modulation, while in c) a Fabry-Perot gauged the diode laser scan. The mode splitting is explained in the text.*

4.4 Mode identification

A microsphere supports many different modes, as can be seen from chapter 2 and from the experimental section 4.3. For the realization of a nano laser or controlled cavity-QED experiments, however, identification of the fundamental mode is crucial. This mode has not only a high-Q, but also the smallest mode volume, providing the strongest field per photon. Since the field outside the microsphere is evanescent (see section 2.2.4), a SNOM operating in the photon tunneling mode is ideally suited to probe and to identify the modes.
A microsphere was mounted in the setup, as described in the previous section, to perform spectroscopy. A near-field probe, placed at shear-force distance (approximately $10\,nm$) to the sphere's surface, scatters photons out of the evanescent field into the fiber, directly coupled to a PMT (see figure 4.6). Thus, moving the tip allows to probe

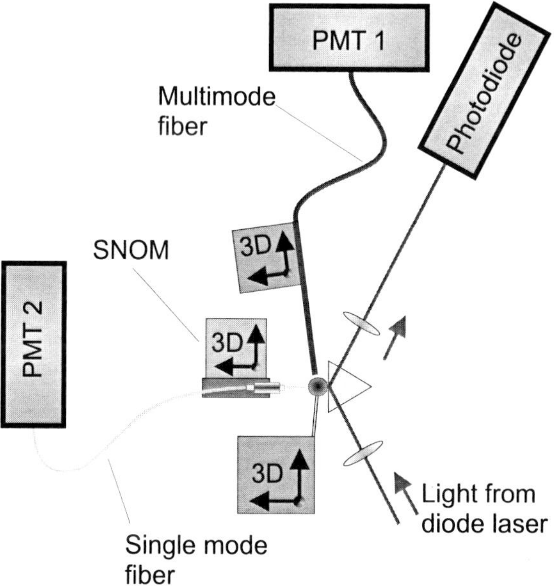

Figure 4.6: *The intensity on the microsphere surface is mapped with the SNOM for mode identification.*

the intensity on the microsphere's surface locally. When the fiber tip is moved in one dimension perpendicular to the sphere equator and the laser is scanned over several modes, a spatio-spectral modemap is obtained. By using the wavemeter, several of these modemaps can be attached until a FSR is covered. Figure 4.7 shows a part of

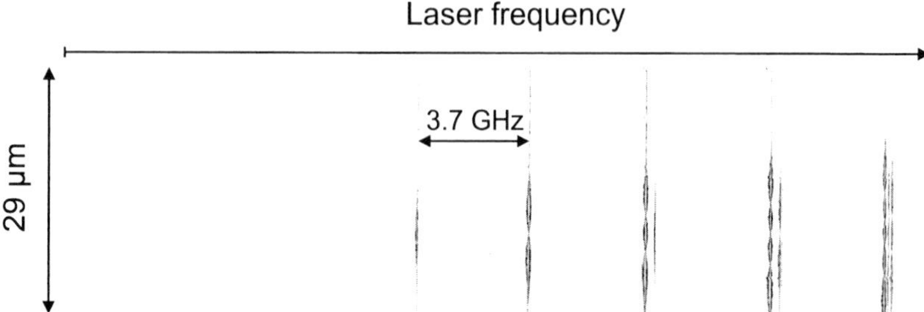

Figure 4.7: *Spatio-spectral mode map of a microsphere with diameter 200 μm. The fainter modes correspond to a higher-order radial mode number n. In the plot the contrast is adjusted such that even modes which are less efficient excited (by a factor of about 100) than the strong low-order n modes are visible [GBS01].*

such a map of a sphere with a diameter of 200 μm. Modes with a different number of lobes corresponding to different values of $l - |m| + 1$ are clearly visible. The spacing of 3.7 GHz between the fundamental mode ($l - |m| + 1 = 1$) having a frequency of 446671.9 GHz and the mode with two lobes ($l - |m| + 1 = 2$) corresponds to an ellipticity of 1 %[1]. This value is derived by putting the numbers into equation (2.55), which shows also that the sphere is an oblate spheroid, because of the higher frequencies of modes with $l - |m| + 1 > 1$ compared to the fundamental one. The reason for this can be found in the production process. Additionally, due to the sensitive detection of the scattered light via the near-field probe, it is also possible to detect much fainter resonances, that can be attributed to modes with higher radial mode number n. These modes are less effectively excited, because the incident angle of the laser was optimized for coupling to lowest-order n modes. In order to visualize these modes, a strongly nonlinear color scale was used in figure 4.7.

The method described above allows a fast screening of a wide spectral range and identification of a particular mode unequivocally. However, for a complete study of a whispering-gallery mode, especially to determine the relative position of the equatorial plane with respect to the horizontal, a two dimensional mapping is necessary, allowing to adjust and to optimize the coupling. In order to map the intensity distribution of an individual mode of the microsphere, the diode laser was locked to a single whispering-gallery mode, using a side band modulation technique similar to the one described in reference [DHK+83]. The SNOM is again used in the photon tunneling mode, while it is scanned across the microsphere's surface. In Figure 4.8 a) the intensity scattered into the fiber is recorded while the tip is scanned, leading to an image of the intensity

[1]The spacing between two adjacent modes depends on the ellipticity. For spheres $\leq 50\,\mu m$ the influence of the stem on the ellipticity usually gets bigger, such that the mode spacing is on the order of 10 GHz.

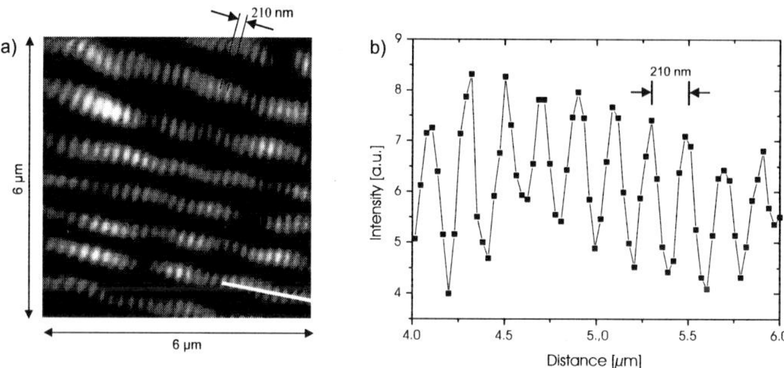

Figure 4.8: *a) Two-dimensional mapping of the intensity distribution of a whispering-gallery mode with the SNOM. During the scan, the laser was locked to this mode. The scanned area is centered vertical around the sphere's equator. b) Cross section of a) marked by the white line. Details are discussed in the text.*

distribution on the microsphere's surface. The scanned area is approximately centered vertically to the equator of the sphere. Several features are clearly distinguished. The various stripes of high intensity are inclined to the horizontal by an angle of about 13°. These rings, parallel to the microsphere's "optical equator" (the plane of the fundamental mode), are a signature of the spatial intensity distribution of a whispering-gallery mode with $l - |m| + 1 \approx 50$. This value was calculated from the distance between neighboring stripes of $1.1\,\mu m$ and the sphere's diameter of $230\,\mu m$, using equations (2.50) and (2.51). The modulation along these stripes is an interference pattern caused by the clockwise and counter-clockwise propagating fields[2]. The distance between two adjacent maxima is expected to be $224\,nm$ given by the mode number m, which is for this mode about 1615 ± 70. The error results from the uncertainty in the measurement of the diameter with an optical microscope. The measured value of $(210 \pm 30)\,nm$ agrees with the theoretical value, determined by 15 measurements on different positions in the image. Figure 4.8 b) shows one of these cross sections.

The observed high modulation depth, together with the small periodicity, clearly demonstrate the very high spatial resolution of the mapping by SNOM to be better than $210\,nm$. Measuring this standing wave pattern allows to estimate the radial mode number n accurately [KDS+96] and thus gives full control over the whispering-gallery modes in terms of n and $l - |m|$.

[2]There seems to be a second slow modulation superimposed along the stripes. But this is a typical example for a SNOM artefact, due to changes in the quality of the shear-force.

4.5 Coupling optimization

The mapping technique introduced above also provides a precise measurement of the orientation of the plane of the fundamental mode (the sphere's "optical equator"), which is a crucial parameter for efficient coupling of the fundamental mode to the prism. The fabrication process of the sphere makes it difficult to tell a priori, how the "optical equator" is exactly oriented. The sphere can for example be slightly tilted on the stem or the elipticity can be perturbed so that "optical equator" and "geometrical equator" (defined by the horizontal plane) do not coincide. As a result the fundamental mode does not couple to the prism. The situation is sketched in figure 4.9. Two rotational degrees of freedom are needed to adjust the "optical equator"

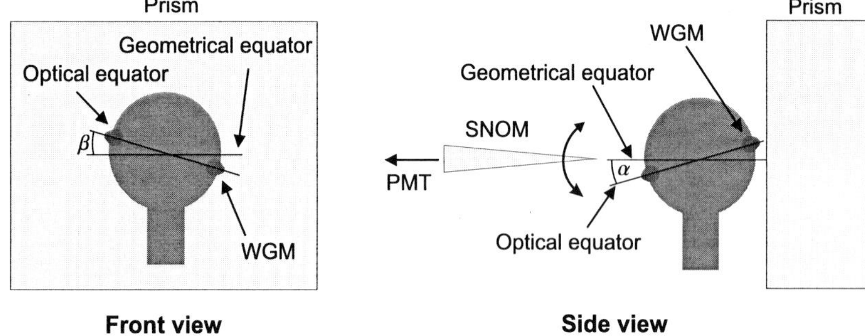

Figure 4.9: *Front and side view of the sphere-prism system. The fundamental mode is inclined due to the fabrication process with respect to the horizontal plane ("geometrical equator") and does therefore not couple to the prism.*

horizontally. The optimization of the inclination angle β is straight forward: in figure 4.8 a) a two dimensional modemap is shown, where the stripes are inclined by an angle of $\beta = 13°$ with respect to the horizontal. By using the goniometer (see section 4.2) this deviation can be corrected right away. For alignment of the second rotational degree of freedom, the sphere is rotated with the rotation stage so that the modes are symmetrically distributed around the horizontal (angle $\alpha = 0$ in figure 4.9). This is done by comparing the mode distribution in direction perpendicular to the horizontal with the topography image, which clearly shows where the modes are located on the sphere.

The exact procedure works as follows: First, a spectrum of whispering-gallery modes with $|m| \approx l$ is taken (figure 4.10 a)). The fundamental mode is extremely weakly coupled to the prism which is hardly seen in the scattering signal (not at all in absorption), but still detectable with the near-field probe. Then the SNOM is used to record simultaneously a 1D topography map perpendicular to the sphere's "geometrical equa-

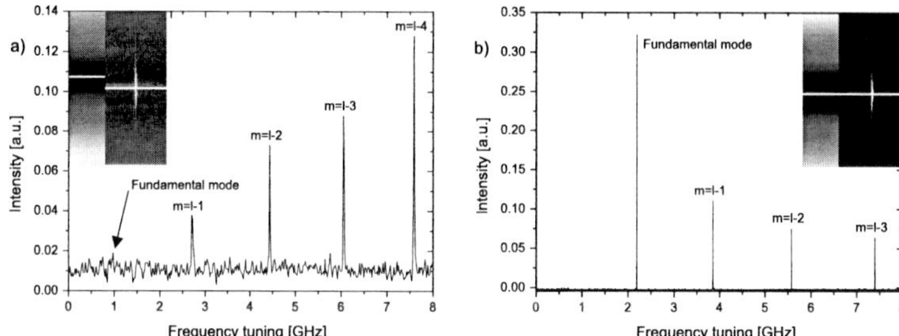

Figure 4.10: *a) A spectrum of modes with $|m| \approx l$ before optimization. The fundamental mode is hardly excited. b) Spectrum of the modes after optimization. The fundamental mode is now coupled strongest. The insets show a 1D topography map, where the color indexes the topography (left) and a spatio-spectral mode map (right).*

tor" and a spatio-spectral mode map . The result is shown as an inset in figure 4.10 a) for the fundamental mode. The white lines indicate the "geometrical equator" (the topography signal is symmetric with respect to it) and the intensity maximum of the fundamental mode (the "optical equator"), respectively. Now, the deviation α, which is the shift of the white lines with respect to each other can be calculated (3° in this case) and adjusted. Figure 4.10 b) shows a spectrum after optimization. The coupling is now optimized for the fundamental mode, which is most strongly excited[3]. Both white lines in the inset in figure 4.10 b) are coaligned. The fundamental mode is now symmetrically centered around the "geometrical equator".
The results shown above clearly demonstrate the power of SNOM for whispering-gallery modes identification, characterization and its use as a tool for optimizing the coupling condition of certain modes in a precise and unique way.

[3]In this measurement the sensitivity of the PMT was reduced, because the sphere was pulled further away from the prism to increase the coupling (see section 2.4). The coupling was also optimized laterally. For this reason, the signal-to-noise ratio in figure 4.10 b) is increased and also the Q is higher.

Chapter 5

Influence of a near-field probe on the Q-factor of a microsphere

In the last chapter it was shown how SNOM techniques can be used for characterizing microsphere resonators. Besides, they open the way to controlled experiments where active material is coupled to high-Q whispering-gallery modes for cavity-QED experiments and the realization of a laser with only one or a few emitters. The ideal strategy for achieving coupling via the evanescent field outside the sphere would be to place the active material directly in the intensity maximum of the fundamental whispering-gallery mode at will. This method of nanomanipulation can be realized by attachment of the active material to the end of the near-field probe [MHMS00, KRMS01]. Unfortunately, as was pointed out in reference [PY99], any material that is used to hold the emitters also couples to the whispering-gallery mode. This coupling introduces an additional loss mechanism through enhanced scattering of photons from the evanescent field. Thus, any influence of a nano-handle, e.g. a fiber tip, has to be minimized. In this chapter, the influence of a near-field probe placed inside the evanescent field of a mode with $m = l$ is studied. It turns out that with a tip diameter of about $100\,nm$ it is possible to maintain a Q-factor exceeding 10^8, even when the tip is as close as $10\,nm$ to the sphere's surface. Finally, a simple model, based on a spherical Rayleigh scatterer modelling the fiber tip is given to provide an easy access to the parameters which are important for the Q-degradation of a microsphere due to a small external scatterer.

5.1 Scanning a fiber tip through low order modes

The mode identification as it is described in section 4.4 was not always carried out with fragile fiber tips with an end diameter of $100\,nm$, but sometimes also with more robust probes having a tip size in the micron range. These tips are much easier to handle, because there is no risk of breaking them during a scan. A zoom into the spatio-spectral mode map clearly shows that micron-sized tips significantly perturb the modes. Figures 5.1 a) and b) show such zooms for a 84 μm sphere for modes with $l-|m|+1=1$ and $l-|m|+1=2$, respectively, which where taken with a tip having a diameter of about one micrometer. A broadening as well as a negative frequency shift

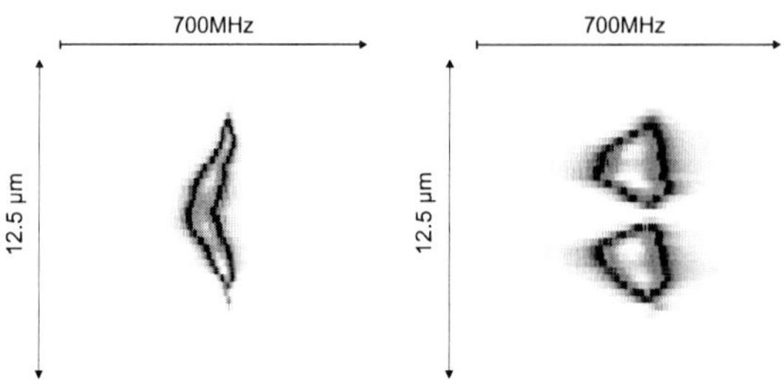

Figure 5.1: *Zoom of a spatio-spectral mode map around a single-lobed mode with $l-|m|+1=1$ (a) and a two-lobed mode with $l-|m|+1=2$ (b).*

are clearly visible as soon as the tip is scanned through the intensity maxima of the modes. For a quantitative measurement of these effects the Q-factor was additionally recorded when the tip and the laser were scanned in a synchronized way (see section 4.4).

In figure 5.2 a)-f) the spectrally integrated intensity distribution, the Q-factor as well as the frequency shift of the two modes as a function of the lateral position are shown. The solid line in figure 5.2 b) shows the theoretically expected intensity distribution (see section 2.3.2) for this mode in a sphere with a diameter of 84 μm[1]. The position dependency of both the Q-factor and the frequency shift resembles the mode structure of the modes with $l-|m|+1=1$ and $l-|m|+1=2$, respectively. Because of the higher initial Q_0 of $2,6\times 10^7$ and a slightly smaller mode volume in the case of the fundamental mode with $l-|m|+1=1$ (figure 5.2 c)), the influence of the tip is much larger than in the case of a mode with 2 lobes (figure 5.2 d)), where the sphere was brought closer to the prism to decrease the initial Q_0 and therefore to decrease the influence of the fiber tip.

The observable lineshift of about 60 MHz in figure 5.2 e) and 5.2 f) is induced by enhanced scattering and dispersive interaction between the fiber tip and the whispering-gallery modes. The fiber tip locally modifies the boundary conditions for total internal reflection of the light inside the microsphere, thus altering the phase condition for resonance [DKL+95]. The extent of this spectral modification is determined by parameters such as the tip's size, shape and scattering cross-section. A quantitative analysis of the experimental data, however, requires advanced numerical calculations which are far beyond the scope of the experiment.

[1] With small fiber tips, which do not perturb the mode, it is also possible to fit the mode with equation 2.49 and let the sphere's diameter (entering the equation with l) as a free fit parameter. In this way mode mapping can be used for a precise measurement of the sphere's diameter.

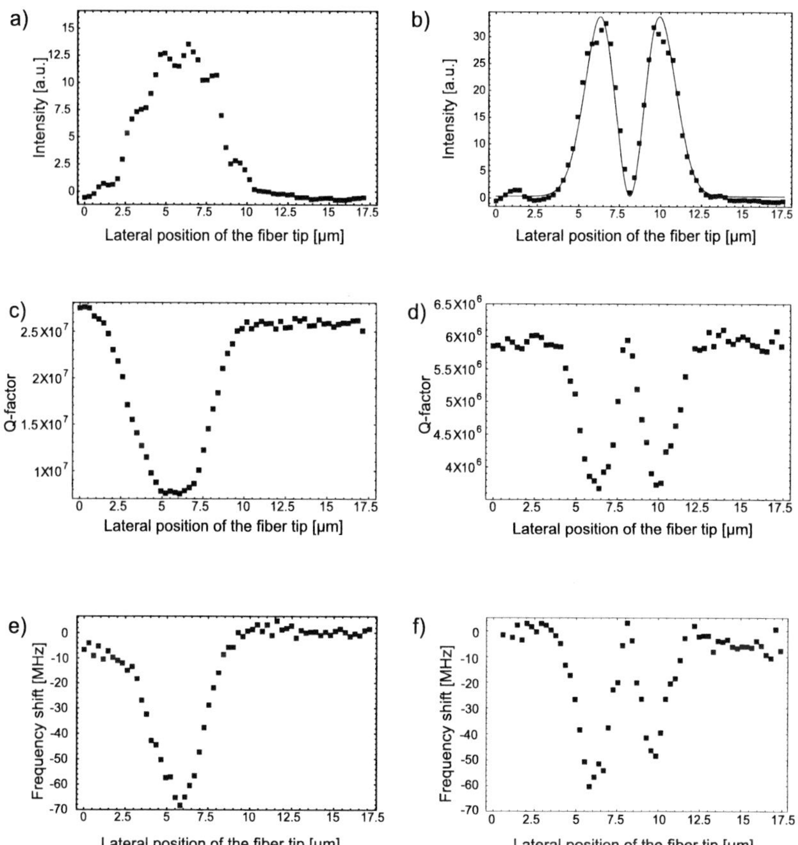

Figure 5.2: *Plot of the spectrally integrated intensity distribution, a) and b), the Q-factor, c) and d), and the line shift, e) and f), versus the lateral position of the fiber tip for a single-lobed mode and a two-lobed mode, respectively. The solid line in b) is a fit with the theoretical expected intensity distribution.*

5.2 Placing a near-field probe in the intensity maximum of the fundamental mode

The last section gave a first impression of the interaction of a fiber tip with a whispering-gallery mode. As one would expect, the effects were strongest when the tip was placed in the intensity maximum of a mode. In this section the influence of a fiber tip, placed in the intensity maximum of the fundamental mode on the Q-factor is studied in more detail. For this most confined mode the Q-reduction due to the scattering by the fiber tip is strongest. This is crucial since the coupling of an active emitter attached to such a fiber tip is also optimal via that mode. The fiber tip perturbs the whispering-gallery

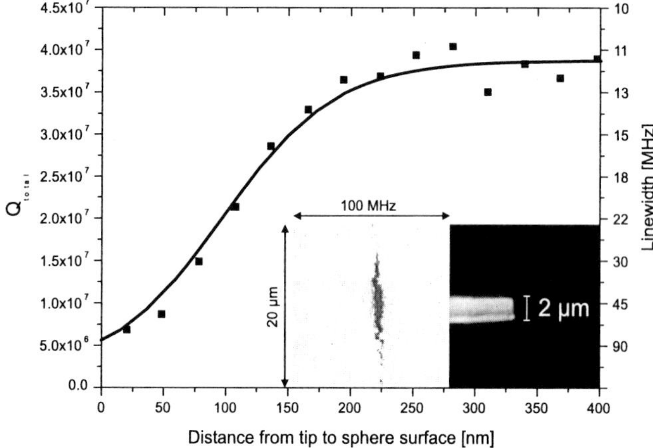

Figure 5.3: *Measured total Q factor, $Q_{total}(z)$, versus the distance z from the tip to the sphere's surface for a $2\,\mu m$ tip. The theoretical curve (solid line) is discussed in the text. The insets show (right) a picture of the fiber tip and (left) spatio-spectral intensity mapping of the mode perpendicular to the sphere's equator (diameter $140\,\mu m$), which displays the characteristic single maximum of the fundamental mode.*

mode leading to a frequency shift as well as the degradation of the Q-factor due to scattering losses (see last section). The latter effect is more pronounced, the larger the tip size relative to the mode volume and the higher the initial Q-factor, Q_0, of the bare sphere-prism system. The following simple relation holds for the measured total Q-factor, Q_{total} [GBS02]:

$$1/Q_{total}(z) = 1/Q_0 + 1/Q_{tip}(z), \tag{5.1}$$

where z is the distance from the tip to the sphere's surface and $Q_{tip}(z)$ is the additional tip-induced Q-factor. The Q-reduction as a function of z for two different tip diameters was measured. First, a larger fiber tip of $2\,\mu m$ diameter (inset in figure 5.3) was placed in the shear-force range which is typically about $10\,nm$ from the sphere's surface and positioned laterally in the maximum of the evanescent field of the fundamental mode by monitoring the intensity collected by the tip. Next, the fiber was pulled back in $30\,nm$ steps using a calibrated piezoelectric element and at every position $Q_{total}(z)$ was derived by measuring the spectral linewidth of the mode. Figure 5.3 shows such an approach curve where the measured Q-factor is plotted versus z. Up to a distance of approximately $200\,nm$ the Q-factor increases roughly exponentially and then saturates at $Q_0 = 3,8 \times 10^7$. The solid line in figure 5.3 is the theoretically expected curve (equation (5.1)), plotted without any fit parameters. $Q_{tip}(z)$ is modelled to be exponentially decreasing, which reflects the approximately exponential decay of the evanescent field outside the sphere. Using equation (2.42) one obtains

$$Q_{tip}(z) = 5.3 \times 10^6 / e^{-2z/r^*}, \tag{5.2}$$

where the factor of 2 in the exponential function takes into account that one has to consider the intensity and not the field.

As pointed out above, Q-reduction due to the approach of a fiber probe is expected to be less pronounced if the initial Q-factor, Q_0, is smaller. To demonstrate this, further approach curves for different Q_0 were recorded and plotted $1/Q_{total}(10\,nm)$ versus $1/Q_0$ (figure 5.4 a)). In this measurement Q_0 was varied over two orders of magnitude by adjusting the sphere-prism gap. For $Q_0 > 5 \times 10^7$ the total Q-factor with the tip approached is dominated by $Q_{tip}(10\,nm) = 5,3 \times 10^6$. This becomes more apparent when plotting $Q_{total}(10\,nm)$ versus Q_0 as shown in figure 5.4 b). The saturation of $Q_{total}(10\,nm)$ at larger Q_0 means that the total Q-factor in such a coupled system is limited by the additional loss mechanism introduced by scattering from the tip. The

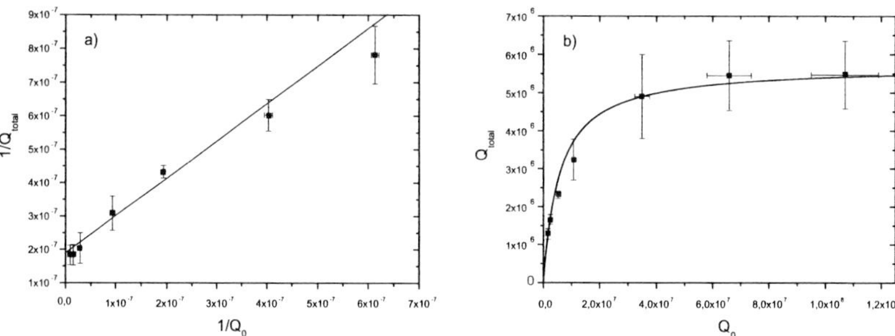

Figure 5.4: *a) Measured inverse total Q-factor, $1/Q_{total}(z)$, with the tip brought to about $10\,nm$ to the sphere's surface versus the inverse unperturbed Q-factor, $1/Q_0$, when the tip is far from the sphere's surface. b) measured total Q-factor, $Q_{total}(z)$, versus Q_0.*

only way to overcome this problem, which is particulary crucial for the very high-Q modes present in microsphere resonators, is to work with a tip that has a very small scattering cross section, e.g. a near-field probe with a very small tip diameter. Figure 5.5 shows the measured approach curve for a tip with a diameter of 80 nm. Within the resolution limit there is no observable Q degradation, although Q_0 is larger than 10^8. Further measurements have proven that this is still valid for 200 nm tips.

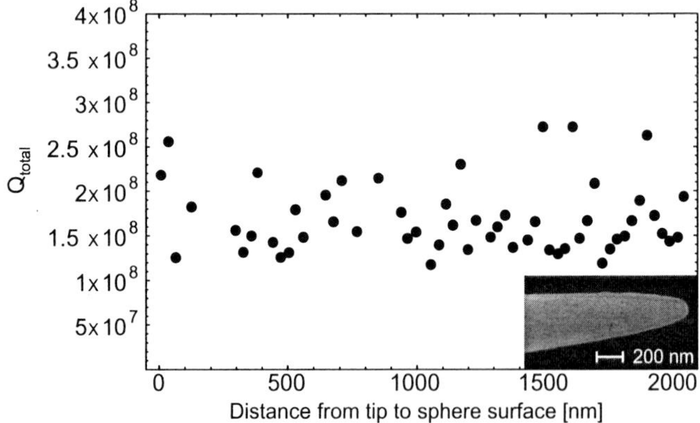

Figure 5.5: *Measured total Q-factor, $Q_{total}(z)$, versus the distance z from the tip to the sphere's surface for an 80 nm tip. The inset shows a scanning-electron microscope picture of the fiber tip.*

In conclusion, it is possible to couple to a fundamental high-Q whispering-gallery mode without degrading the Q-factor via a sharp optical fiber tip. Such a tip could act as a nanotool in order to establish a well controlled coupling of a nanoscopic amount of active material to a high-Q mode of a microsphere resonator in experiments which aim at the realization of cavity-QED experiments or novel nanoscopic light sources.

5.3 A simple model

This section gives a deeper insight into the parameters which are important for the Q spoiling of a microsphere due to the presence of a near-field probe. A model is developed to get a qualitative understanding for the order of magnitude and for the dependencies of the quantities involved in the interaction of the tip with the mode[2]. The only underlying process to be considered is Rayleigh scattering, which restricts the model to tips with an end diameter, which is small compared with the wavelength of the mode. As a further simplification, the tip is modelled as a spherical scatterer

[2]A more exact model would require exact knowledge of the tip's geometry and heavy numerics.

with radius r (see figure 5.6). Also, in the mode the incident field on the scatterer is

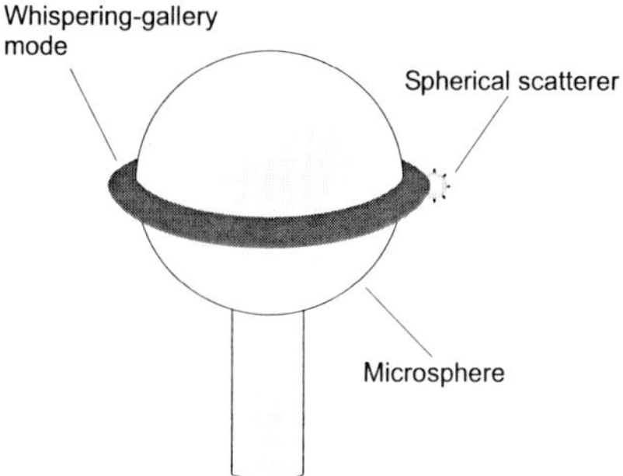

Figure 5.6: *A spherical scatterer in the evanescent field of a whispering-gallery mode.*

assumed to be a plane monochromatic wave, while in the real experiment the scatterer is placed in an evanescent field. The *total scattering cross section* σ is then given by [Jac98]:

$$\sigma = \frac{8\pi}{3} k^4 r^6 \left|\frac{\epsilon - 1}{\epsilon + 2}\right|^2, \tag{5.3}$$

where k is the wave number and ϵ is the dielectric constant, which is for silica glass 2.13. Notice the variation of the total scattering cross section on the wave number as k^4 (or in wavelength as λ^{-4} known as Rayleigh's law) and the strong r^6 dependency on the radius of the scattering particle. The difference in the total scattering cross section for a particle with a radius of $100\,nm$ to a particle with a radius of $50\,nm$ is about two orders of magnitude.

Let's assume a fundamental mode with a mode volume V_{mode} according to section 2.3.4 in a silica microsphere with radius a. The light is travelling in a great circle around the sphere's perimeter losing the energy ΔE in every round trip, when passing the scatterer. The loss is given by the following ratio of the areas:

$$\frac{\Delta E}{E_0} = \frac{\sigma \frac{I(d)}{I_0}}{A_{mode}} \tag{5.4}$$

where E_0 is the energy stored in the resonator, A_{mode} is the cross section through the mode volume, i.e. $A_{mode} = \frac{V_{mode}}{2\pi a}$. $I(d)$ denotes the radial intensity distribution which can be derived from the radial field distribution, already discussed in section 2.2.4. $\frac{I(d)}{I_0}$ is then the normalized intensity distribution, which weights the total scattering cross section, since the losses will be less in regions where the intensity is lower, which means that this factor accounts for the intensity distribution of the mode. Equation 5.4 is only true, if the total scattering cross section is small compared to A_{mode}. As the scatterer is only outside the sphere, $I(d)$ is given by the exponentially decaying evanescent field described by equation (2.42)

$$I(d) = |\mathbf{E}(d)|^2 \propto e^{-2d/r^*}. \tag{5.5}$$

Because of the scattering, the energy in the resonator will be decreased after one round trip to a value $E_{rt} = E_0 - \Delta E$. Division by E_0 and using equations (5.4) and (5.5) yields

$$\frac{E_{rt}}{E_0} = 1 - \frac{\Delta E}{E_0} = 1 - \frac{\sigma}{A_{mode}} e^{-2d/r^*}. \tag{5.6}$$

The energy inside the resonator $E(t)^*$ is proportional to the the square of the field inside the cavity. Inserting (2.7) into equation (2.3) shows an exponential decay with the decay time τ for the field. Thus the temporal evolution of $E(t)^*$ can be written as[3]:

$$E(t)^* = E_0 e^{-t/\tau}. \tag{5.7}$$

The decay time is given by

$$\frac{1}{\tau} = \frac{1}{\tau_0} + \frac{1}{\tau_{sct}} \tag{5.8}$$

since $\tau \propto Q$ and the different contributions to the total Q sum up reciprocally. τ_0 describes the decay time of the undisturbed cavity, whereas τ_{sct} denotes a decay time due to additional losses introduced by the scatterer. The time for one round trip t_{rt} can easily be calculated by

$$t_{rt} = \frac{2\pi a N}{c}, \tag{5.9}$$

where c denotes the speed of light. By putting equation (5.8) into equation (5.7), one can rewrite the left hand side of equation (5.6) as

$$\frac{E_{RT}}{E_0} = e^{-t_{rt}/\tau_0} e^{-t_{rt}/\tau_{sct}}. \tag{5.10}$$

Combining equations (5.6) and (5.10) results in

$$e^{-t_{rt}/\tau_0} e^{-t_{rt}/\tau_{sct}} = 1 - \frac{\sigma}{A_{mode}} e^{-2d/r^*}, \tag{5.11}$$

where the first factor of the left hand side can be set to one, since t_{rt} is orders of magnitude smaller than τ_0 in the case of a high-Q silica microsphere. This allows to rewrite equation (5.11) and to calculate the decay constant τ_{sct}

$$\tau_{sct} = \frac{t_{rt}}{-\ln(1 - \frac{\sigma}{A_{mode}} e^{-2d/r^*})}. \tag{5.12}$$

With the knowledge of τ_{sct} one can finally calculate the total Q-factor Q_{total} according to equation (2.8) to be

$$\frac{1}{Q_{total}} = \frac{1}{Q_0} + \frac{1}{Q_{sct}}, \tag{5.13}$$

[3]In equation (2.3) E(t) denotes a field amplitude, whereas in equation (5.7) $E(t)^*$ denotes the energy inside the resonator, which is proportional to the intensity. Therefore the factor 2 in the exponential vanishes.

with $Q_{sct} = \omega_0 \tau_{sct}$ (see equation (2.7)). Inserting τ_{sct} into equation (5.13) leads to the desired expression:

$$Q_{total} = \frac{1}{\frac{1}{Q_0} - \frac{\ln(1 - \frac{\sigma}{A_{mode}} e^{-2d/r^*})}{t_{rt}\omega_0}}. \tag{5.14}$$

Equation (5.14) is plotted in figure 5.7 for different initial Q-factors Q_0. The scatterer's

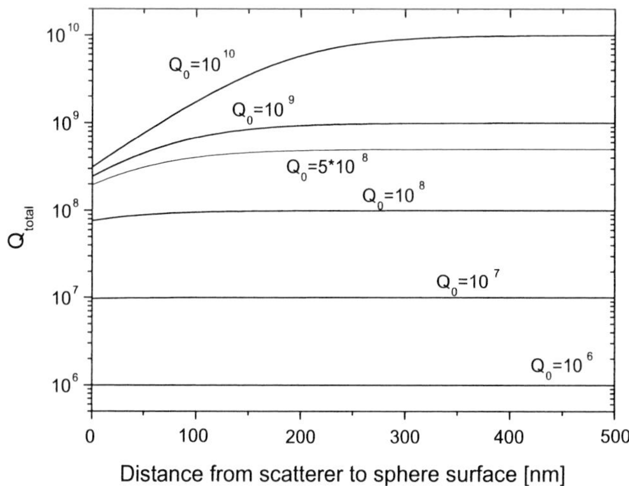

Figure 5.7: *Total Q-factor Q_{total} versus the distance of the scatterer to the microsphere's surface for different initial Q-factors Q_0. The radius of the scatterer was set to be $50\,nm$ and the diameter of the sphere $100\,\mu m$.*

radius was set to be $50\,nm$, the sphere's diameter was $100\,\mu m$ having a resonance at $670\,nm$ ($\omega_0 = 2.8 \times 10^{15}\,rad/s$). Both parameters enter the calculation of the mode volume (see equation (2.56)). Figure 5.7 shows that a fiber tip with an end diameter of about $100\,nm$ has a negligible influence on the Q-factor of a microsphere with a diameter of about $100\,\mu m$ and Q_0 of 10^8 [4]. The predictions of the model agree therefore quite well with the experimentally obtained data for the $80\,nm$ tip. The plots also show a kind of saturation behavior for $Q_0 > 10^8$. The scatterer restricts the total Q to about 2×10^8. Also this saturation was nicely demonstrated with a $2\,\mu m$ tip in section 5.2, where the tip gave an upper limit of $5,3 \times 10^6$ for Q_{total}. However, both cases cannot be directly compared, since such a big tip cannot be treated as a Rayleigh scatterer anymore.

[4]Distance zero means that scatterer and sphere surface are in contact.

The other important question in this context is the influence of the size of the scattering particle on the Q-factor. In figure 5.8, function (5.14) is plotted for different scatterer sizes. The sphere has a Q_0 of 10^8, the other parameters are the same as above. Comparing the results with experimental data shows that the model is too

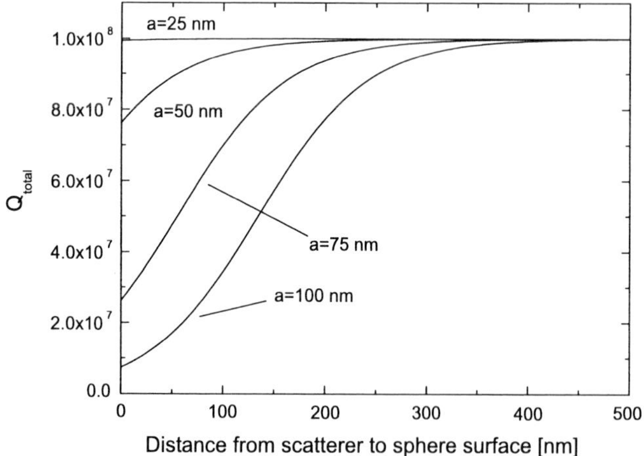

Figure 5.8: *Total Q-factor Q_{total} versus the distance of the scatterer to the microsphere's surface for different scatterer sizes. The sphere's diameter was set to be $100\,\mu m$ and having an initial Q-factor Q_0 of 10^8.*

pessimistic. A tip with an end diameter of $200\,nm$, placed in the intensity maximum of a microsphere with a diameter of $200\,\mu m$ did not spoil the Q, even when Q_0 was higher than 10^8. The plot shows also that the Q-factor starts to decrease at further distances with increasing scatterer size. This effect could not be confirmed experimentally with tips, which could be modelled by a spherical Rayleigh scatterer, because they did not show any measurable Q-degradation. However, in experiments with tips having a size of more than $10\,\mu m$ acting on modes with $l > m$, the Q-factor started to decrease at a distance of about $500\,nm$ [GDBS01], whereas for tips smaller or equal to $2\,\mu m$ the degradation started only at $200\,nm$.

To summarize, the model shows all experimentally observed effects, but overestimates the Q-degradation due to tip induced scattering. A reason for this, is for example the assumption that the incident field on the scatterer is simply modelled by a plane wave. In order to demonstrate a Q-spoiling caused by the presence of a near-field probe, micron sized tips needed to be used, whereas the model predicts this already for a $r = 50\,nm$ tip and $Q_0 = 10^8$. But the model confirms that tip induced scattering results in an upper limit for Q_{total}.

Chapter 6

Controlled coupling of fluorescent nanoparticles

In this chapter, the different concepts and techniques which were previously presented are combined to perform controlled experiments, in which nanoparticles are coupled to high-Q microsphere resonators. Two basic strategies are pursued. In the first one, the microsphere is coated with the nanoparticles and the use of the confocal microscope allows to select and to address one of the particles on the sphere's surface. With this technique, the coupling of even a single nanocrystal could be demonstrated. However, in this approach the particles are randomly distributed due to the coating process and therefore there is no possibility to control and optimize the coupling conditions. In other words, one relies on trial and error to obtain efficient coupling between one single nanoparticle and a whispering-gallery mode. The second method, however, gives full control over the coupling condition. Here, a single nanoparticle is attached to the end of a near-field probe, which allows to place it at will at any position on the microsphere's surface. With this technique, a more detailed investigation of the coupling is performed.

6.1 Individual nanoparticles addressed on a microsphere by means of confocal microscopy

The coupling of active material to high-Q microsphere resonators was already previously demonstrated in various ways [VFG+98, FLW99, FPL+00], even lasing was observed [STH+96, CMA+01]. The peculiarity of the following experiments is the coupling of a single nanoparticle to the microsphere. The individual particle can be identified on the sphere's surface and its location with respect to the whispering-gallery modes can be verified. For these experiments, the microsphere spectroscopy unit (section 4.2) is combined with the beam scanning confocal microscope (see section 3.2).
Before the microsphere was installed into the setup, it was dipped for a while in a solution containing a certain concentration of nanoparticles, depending on the desired density of particles on the sphere's surface. Great care has to be taken when choosing

the solvent for the nanoparticles, since many solvents either spoil the Q-factor substantially or do not produce a good film on the sphere. It turned out that depending on the solubility of the particles, methanol or toluene met the requirements. These solvents did not cause any significant Q spoiling. After the dipping procedure, the microsphere was covered with some nanoparticles , and was then installed in the spectroscopy unit. The Q-factor was measured and the coupling adjusted as already described in section 4.3. By varying the sphere-prism gap, the Q-factor, but also the out-coupling efficiency, can be adjusted. Spectroscopy of the microsphere was performed before and after the experiment to check the coupling. Then, the beam scanning confocal microscope was moved on its rail towards the spectroscopy unit in order to excite the nanoparticles on the surface of the sphere. The complete experimental setup is shown in figure 6.1.
The microscope was first used in the wide field imaging mode to illuminate a wider area on the sphere. Fluorescence of the nanoparticles was collected with the microscope objective. Before being sent to the ultrasensitive CCD-camera (Hamamatsu ORCA ER), the fluorescence was separated from the excitation light by a dichroic mirror (DCM), a notch filter (NF) and a color glass filter. The wide field mode allows to identify the particles and their position on the sphere. Since the sphere is imaged on the CCD, when the spectral filters are removed from the detection path, one also knows exactly where the nanoparticles are located with respect to the sphere's geometry. This allows to choose particles close to the equatorial region. By switching to the confocal mode one can selectively excite the chosen particle on the microsphere's surface. By flipping down some mirrors (FM), the collected light can be sent to a spectrometer or avalanche photodiodes (APD 1 and APD 2[1]). Light, which is emitted by the nanoparticle and which coupled to the whispering-gallery modes, is extracted out of the sphere by the prism coupler, which serves as an out coupling port. Here also, flip mirrors are used to send the light to the CCD, the spectrometer or APD 2. With the setup, one can observe the particle in two different ways: either one observes the emission into free space or the emission into the whispering-gallery modes. When APD 1 is combined with any of the other detectors, both emission paths can be recorded simultaneously. Switching to another particle is done with the computer, which controls the galvo-drives of the beam scanning confocal microscope. Therefore, there is no need for an additional alignment. Other computers read the signals from the various detectors, if necessary synchronized by a LAB-VIEW software.

[1]APD 2 was sometimes replaced by a single photon counting photomultiplier tube, when ultimate sensitivity was not required.

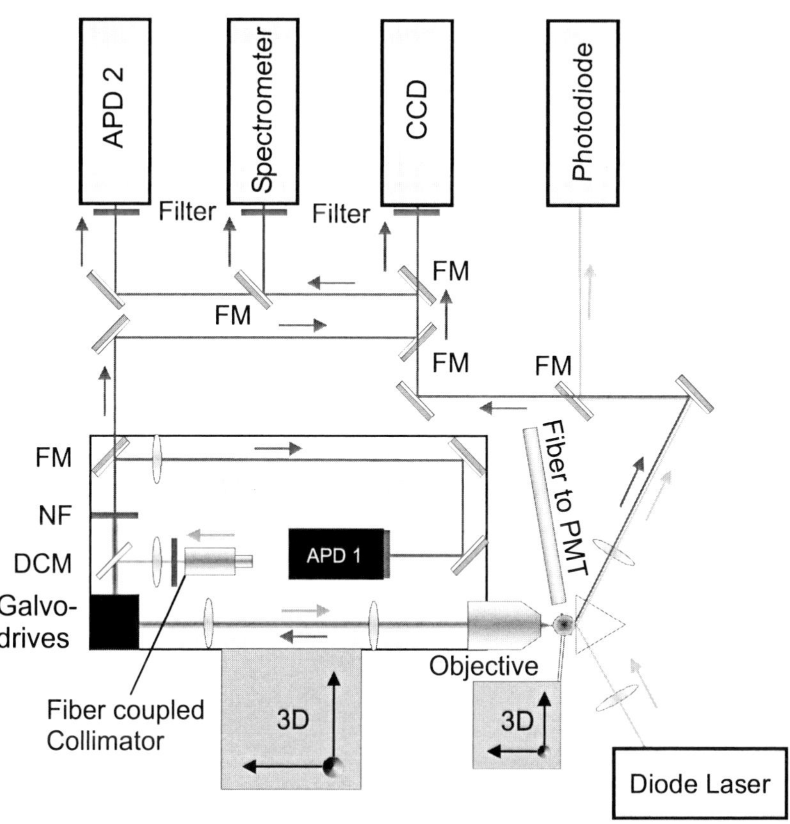

Figure 6.1: *Experimental setup for microscopy and spectroscopy of nanoparticles on a high-Q silica microsphere. Light from the diode laser is coupled into the microsphere resonator to control its coupling to the prism. The beam scanning confocal microscope is used to address and to detect single nanoemitters on the microsphere's surface. The fluorescence light is collected and sent to various detectors via flip mirrors (FM). Light which couples to the resonator's modes can be extracted via the prism coupler and can also be sent to the detectors. Dichroic mirror (DCM), notch filter (NF), avalanche photodiode (APD), photomultiplier tube (PMT).*

6.1.1 Dye-doped beads

Although the setup has single molecule detection sensitivity, first experiments were performed with dye-doped beads, because they provide a stronger fluorescence signal. Since many or at least a few molecules participate in the fluorescence from a single bead, the problem of photobleaching is relaxed and more complex experiments, which require longer observation times of a single bead, are feasible. The size of commercially available beads ranges from several microns down to $20\,nm$.

In the experiment, sub-wavelength polystyrene spheres doped with dyes[2] were used, ranging in size between $100\,nm$ and $500\,nm$. These beads were suspended in methanol to coat the sphere. Usually, there were about 20 beads prepared on the microsphere's surface, a value that was obtained by wide field imaging of the sphere's surface. By focussing through the microsphere, it was even possible to image beads which were on the opposite side of the sphere. Thus, one could count quite accurately the number of beads. Q-factors $> 10^8$ could still be measured with this number of beads on the surface. As already described above, it is possible to address only a single nanoparticle, in this case a single bead, with the excitation laser coupled into the beam scanning confocal microscope. The fluorescence from this single nanoscopic emitter can then be detected via the microscope objective or via the prism output coupler.

In order to test and to demonstrate the unique feature of this method, the following measurement was performed: first, a particular bead was selected in the wide field image (see inset figure 6.2). Then this bead was addressed with the microscope in the confocal mode. Figure 6.2 shows the signal that was detected via the prism output coupler when the focus of the excitation laser beam was moved several micrometers away from its initial position. Clearly, the signal drops rapidly to the background level. This ensures that indeed only the selected bead was excited. It should be mentioned that one cannot expect to work with a diffraction limited spot on the sphere's surface, because the sphere is a three dimensional object and the bead will always be located a little bit off-center. Therefore, a part of the beam is refracted by the sphere when it is focussed on the bead. This effect is stronger the further the bead is located away from the microscope objective. Then, as a problem there is the possibility that by chance another bead is also excited, much weaker but still significantly. Therefore, this needed to be checked at the beginning of every experiment.

In order to see how the coupling of a nanoparticle manifests in the spectrum of the fluorescence, a single bead with a size of $190\,nm$ was excited. Its fluorescence was first detected via the microscope, then via the prism coupler. In such experiments, the bead was usually excited with a power of $18\,\mu W$, which corresponds to a power density of about $9\,\frac{kW}{cm^2}$ in the focus. The spectrum of the bead detected, via the microscope, can be seen in figure 6.3 a), while in b) the spectrum recorded via the prism is plotted. In both cases a grating with $600\,groves/mm$ was used resulting in a resolution

[2]Unfortunately, the companies sometimes do not provide full information about the dyes, which are actually infiltrated into the beads, since this is usually confidential.

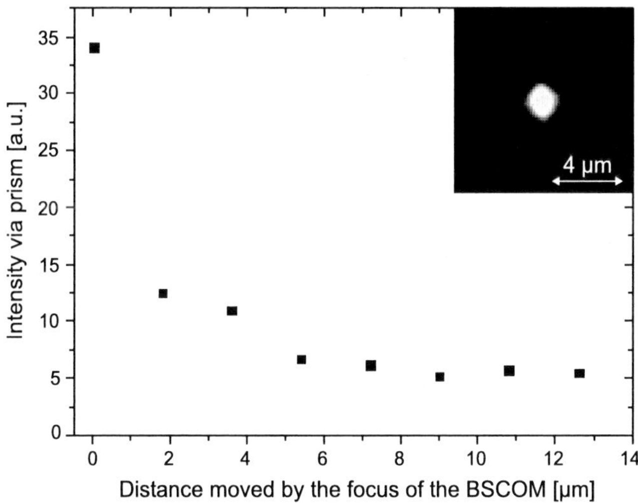

Figure 6.2: *Addressing a single 500 nm dye-doped bead on the microsphere. The intensity of the fluorescence from the bead was detected via the prism coupler. Initially, the laser beam was focused on the bead using excitation via the beam scanning confocal microscope (BSCOM). Then, the beam was moved away by the indicated distance. The inset shows a CCD image of the bead on the sphere's surface [GdSMM⁺03].*

of 0.2 nm (slitwidth 100 μm). The signal detected via the prism was for this bead size about 50 times weaker than the signal detected with the confocal microscope. This was compensated with an integration time of 30 s in the case of the prism output coupler, while the spectrum was only integrated for about 0.5 s when the microscope was used. Figure 6.3 a) shows the broad spectrum of dye molecules at room temperature, without any special features, similar to what one expects from a bead on a coverslip. In the case of the fluorescence extracted out of the sphere, however, a very pronounced modulation of almost 100 % is observed. The envelope of the spectrum is the same as in figure 6.3 a). A zoom in the measured spectrum (grating with 1800 $groves/mm$, resolution 0.07 nm) is depicted in figure 6.3 c), which shows the regular modulation in detail. The main feature is a periodic double peaked structure, which originates from free spectral range (FSR) of the cavity modes in TE and TM polarization. This assertion is validated by deriving the FSR of the microsphere from the measured spectrum, which was before converted into the frequency space, by using Fast Fourier Transform (FFT). The FFT yields a spacing of 970 GHz for two consecutive peaks of the same

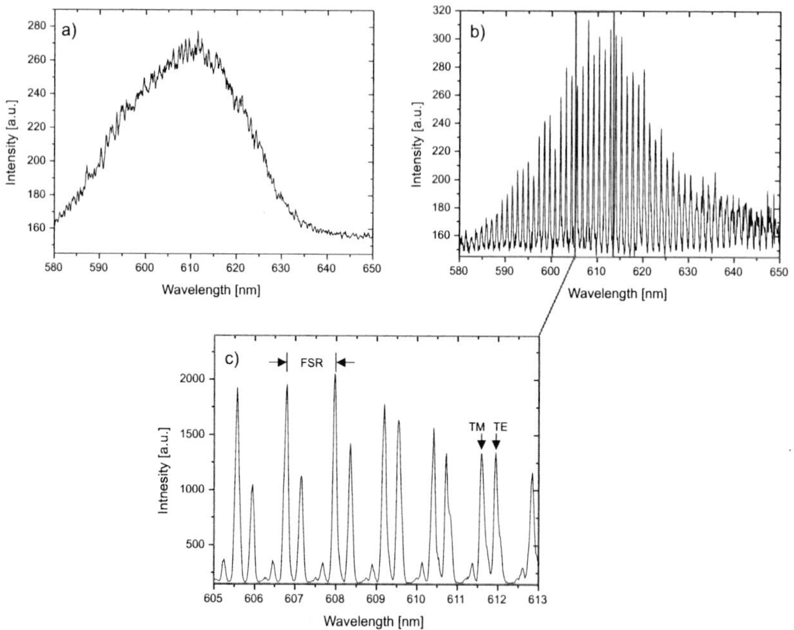

Figure 6.3: *a) Spectrum of a single 190 nm bead recorded with the confocal microscope. b) Spectrum of the fluorescence of the same bead measured via the prism coupler. c) High resolution spectrum of the bead detected via the prism. Arrows denote the TE- and TM- modes and the free spectral range (FSR), respectively.*

polarization, which corresponds to a spacing of $1.2\,nm$ at a wavelength of $609\,nm$. The theoretically expected FSR (see equation (2.53)) for the sphere which had a diameter of $73\,\mu m$, is $(895 \pm 90)\,GHz$. The spacing should therefore be $(1.1 \pm 0.1)\,nm$ around $609\,nm$. The error arises from the measurement of the diameter with the long distance microscope. The assumed accuracy is $4\,\mu m$. Both values agree within the accuracy of the measurement, whereas the resolution of the spectrometer does not need to be considered. Comparison with the theory could also help to identify the peaks arising from TE or TM modes (see equation (2.54))[3]. The expected spacing of $0.88\,nm$ (at $609\,nm$) was clearly present in the FFT. The most pronounced peaks in figure 6.3 c) can be attributed to first order ($n = 1$) radial order modes. The fainter peak in be-

[3]This was additionally confirmed by analyzing the fluorescence with a polarizer inserted in front of the spectrometer.

tween originates from second order ($n = 2$) radial order modes. A second faint peak seems to be missing, but in fact it is superimposed to the strong first order radial mode of the other polarization. Measurements on spheres with other diameters have shown both faint ($n = 2$) peaks. Due to the finite resolution of the spectrometer, modes with the same radial, but different quantum number m, could not be resolved.

An additional proof that the prism serves as an output port of the coupled light is the fact that the fluorescence signal detected via the prism completely disappears when the sphere is moved by only about 100 nm away from the prism, out of its coupling region. This is done by monitoring the coupling of the diode laser to the sphere with the multimode fiber until no coupling is observable anymore. One important question has not yet been addressed, namely the coupling dependency of the bead on its position on the microsphere. One expects that the coupling to modes differing in m should vary as well as the coupling of these modes to the prism coupler. This effect was indeed observed in the experiment: For beads in the equatorial region of the sphere the coupling could be observed (as in the example shown above), while beads which where located closer to the poles did not show any signal when detected via the prism. The situation is sketched in figure 6.4. It should be stressed that here, coupling refers to the combined effect of coupling into the sphere and out of the prism, because the prism coupler has a mode selectivity which was already discussed in section 4.3. It is evident that the bead couples to many modes in the sphere, but these modes do not necessarily couple to the prism. More quantitative measurements where the bead is moved along the sphere and thus the coupling to certain modes is modified will be discussed in section 6.2.

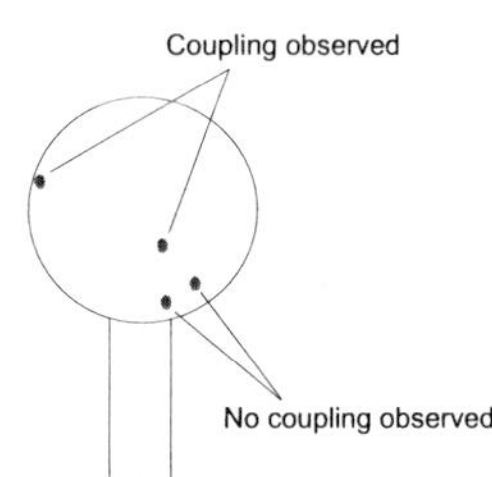

Figure 6.4: *Dependency of the coupling on bead's position in the sphere prism system.*

The experiment clearly demonstrates the coupling of a single nanoparticle to high-Q whispering-gallery modes. Qualitatively the same results were obtained with a single 100 nm bead. Hence, one can deduce that even the coupling of beads with sizes below 100 nm should be detectable. The signal-to-noise ratio will become worse, but the experiments will not show any new results, since one is still coupling an ensemble of molecules to the microsphere. A real step further would be the coupling of a real quantum emitter, like a single molecule or a nanocrystal. This problem is tackled in the next section.

6.1.2 CdSe/ZnS nanocrystals

Nanocrystals and some of their properties were already briefly introduced in section 3.2.4, where they were used to test the confocal microscope. Here, the important properties of these particles for this work are summarized. In order to explore deeper the subject of the optical properties of single nanocrystals, references [EB99a] and [ENSB99] are recommended as a starting point.
The nanocrystals used in this work are quantum dots synthesized as colloids [MNB93, GTR+02, RTS+02]. They consist of a CdSe core, which is overcoated by a ZnS capping layer. Its most apparent effect is an increase in the fluorescence quantum yield. Values as high as 50 % have been reported at room temperature [DRVM+97]. The size of these nanoparticles is tunable during the synthesis ($1.5 - 100\,\mathring{A}$) whereas the size distribution can be smaller than $5\,\%\,rms$. In this size range, the nanocrystals are smaller than the diameter of the bulk Bohr exciton ($112\,\mathring{A}$ for CdSe). As a result, the electronic structure is dominated by quantum confinement effects in all three dimensions, making these nanocrystals truly zero-dimensional structures [ENSB99]. The quantum confinement is responsible for the optical properties. Band edge emission and absorption wavelengths can be tuned across the visible range and into the near infrared by controlling the size of the nanocrystals. In addition they have a spectrum of discrete, atomic-like energy states [NERB96]; for this reason they are sometimes also called artificial atoms. Photon correlation measurements have shown that a single nanocrystal emits antibunched light [MIM+00, LBG+00, MHG+01] and can thus be treated as a quantum emitter, such as a single molecule or an atom.
In contrast to the case of well isolated single atoms the transition linewidth of nanocrystals is rather broad. Variations in size and shape result in an extensive inhomogeneous broadening of the ensemble spectrum. Values of about $28\,nm$ have been reported [ENSB99, SBPM02]. However, even in the case of a single nanocrystal the spectrum is still inhomogeneously broadened, due to spectral diffusion. This effect together with phonon interaction leads to a spectral width of about $15\,nm$ at room temperature [SBPM02]. At cryogenic temperatures (10 K), spectral diffusion leads to a width of about $0.3\,nm$ on the time scale of fluorescence measurements [EB99b], a value which can serve only as an upper bound for the true intrinsic width.
It was shown that spectral diffusion is correlated with fluorescence intermittency [EB99b, NSW+00], also called "blinking". This turning "on" and "off" behavior has been interpreted in terms of an Auger ionization model. In this model, the off periods are the times when the dot is ionized and the fluorescence is quenched by nonradiative Auger recombination [ER97]. Excitation light creates electron-hole pairs which recombine radiatively, emitting light until the dot is ionized, either thermally or due to Auger autoionization. The ejected electron (hole) is now trapped in the surrounding matrix, initiating the off-period, where nonradiative Auger processes in the charged crystal quench all the fluorescence. The ejected electron (hole) relaxes after a certain time, initiating an on-period where the nanocrystal fluoresces again.

This binary fluorescence blinking, rather than a stepwise or continuous dimming of the emission, is strong evidence that the fluorescence originates from a single nanocrystal [ENSB99].

6.1.2.1 Ensemble measurements

In order to couple nanocrystals to a microsphere, the sphere was dipped in a solution of nanocrystals diluted in toluene. This solvent does not spoil the Q-factor, so that it is possible to maintain a Q-factor as high as 10^9, which corresponds to a Finesse of 10^6 for typical sphere sizes of $100\,\mu m$. The concentration of nanocrystals on the microsphere's surface is then about 0.5 nanocrystals per μm^2 (see inset in figure 6.5 b)). It was also attempted to increase the stability of the nanocrystals by adding $40\,\mu l$ of polymethylmethacrylate (PMMA) dissolved in chloroform (10 wt.-%) to $3\,ml$ of the toluene-nanocrystal solution. However, the price for the increased stability of the emitters was a substantial Q-spoiling of the sphere. The obtainable Q-factors were in this case on the order of several times 10^6. Therefore, most of the experiments with nanocrystals were performed without PMMA. The ensemble measurements were performed by wide field illumination (power density $\approx 80\,W/cm^2$) of a sphere coated with the nanocrystals.
The spectrum of CdSe/ZnS nanocrystals emitting at $620\,nm$ detected via the microscope is shown in Figure 6.5 a). It shows the typical inhomogeneously broadened spectrum of nanocrystals, described above, with a width of $23\,nm$. The fluorescence signal detected, via the prism outcoupler, shows a 100 % modulation (figure 6.5 b)), as was seen already in section 6.1.1 for a single bead. Here also the periodic double peak structure is clearly observable, when zooming into a region of the spectrum as is done in figure 6.5 c). Using the same FFT analysis as in the case of the bead leads again to an unequivocal identification of the FSR of the sphere. The FFT resulted in a value of $427\,GHz$, corresponding to a spacing between two peaks of the same polarization and radial mode number of $0.56\,nm$ at $629\,nm$. This value is in good agreement with the calculated FSR of the sphere ($(422 \pm 11)\,GHz$), which had a diameter of $155\,\mu m$. The TE and TM modes could also be identified with the FFT by their splitting. The weaker peaks are again attributed to modes with radial mode number $n = 2$.
The identification of the TE/TM modes was also done with an alternative method to FFT. A whole set of spectra was measured with a polarizer in front of the spectrometer, while the polarizer was rotated in $10°$ steps. In figure 6.6 the intensity of a TE and a TM mode is plotted for different polarizer angles. A clear anti-correlation is observable.

The results shown here are in good agreement with the measurements performed with the dye-doped beads. But the nanocrystals allow, in contrast to the beads, to perform measurements with single quantum emitters. Single molecules would be in principle an alternative option, but some test experiments with dye molecules (Alexa Fluor 568) have shown the superiority of the nanocrystals as far as photostability is concerned.

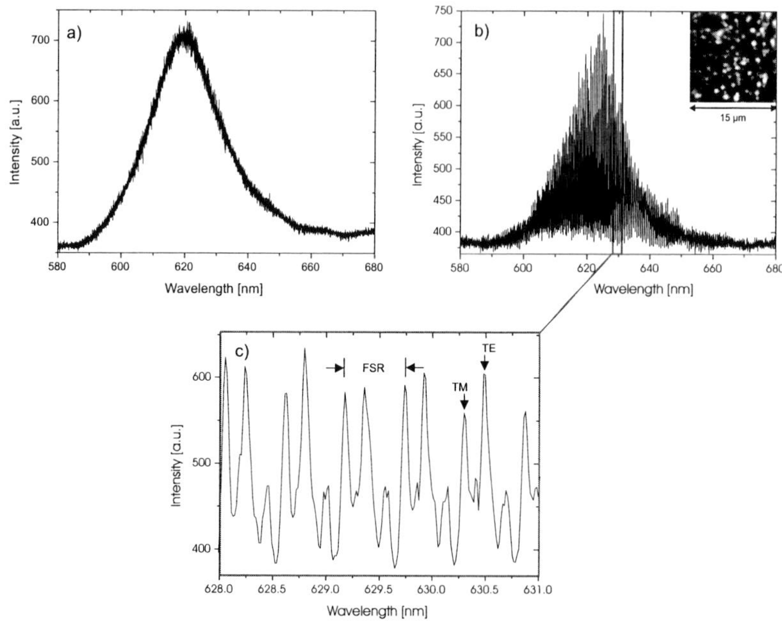

Figure 6.5: *a) Fluorescence spectrum from nanocrystals on a microsphere surface detected via the microscope. b) Fluorescence spectrum from the same nanocrystals coupled out of the microsphere with the prism coupler. The inset shows a CCD-camera image of the nanocrystals on the sphere's surface. With such concentrations, Q-factors in the order of* 10^9 *could still be maintained. c) Zoom in the measured fluorescence spectrum. The vertical arrows indicate TE and TM modes.*

6.1.2.2 Measurements on single nanocrystals

As in the case of the beads, the microscope allows to select a single nanocrystal on the microsphere's surface. From imaging without notch and longpass filters, one also knows exactly where the emitter is located on the sphere. On the left hand side in figure 6.7, some nanocrystals on a sphere are imaged, while the microscope is used in the wide field mode. The particle in the circle was then selected at will to be excited by the microscope in the confocal mode. The result can be seen on the right hand side of figure 6.7, where only the fluorescence of the chosen particle is now observed. This allows to perform experiments on the single nanocrystal level on the surface of a high-Q microsphere resonator, while the sphere is coupled at the same time to the prism

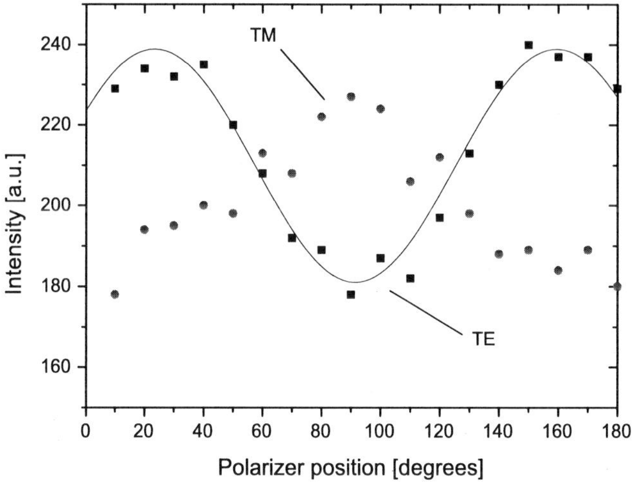

Figure 6.6: *Dependance of the peak height for TE (squares) and TM (circles) modes as a function of polarizer position in front of the spectrometer. A sinusoidal fit is plotted for the TE mode to guide the eye [GdSMB+04].*

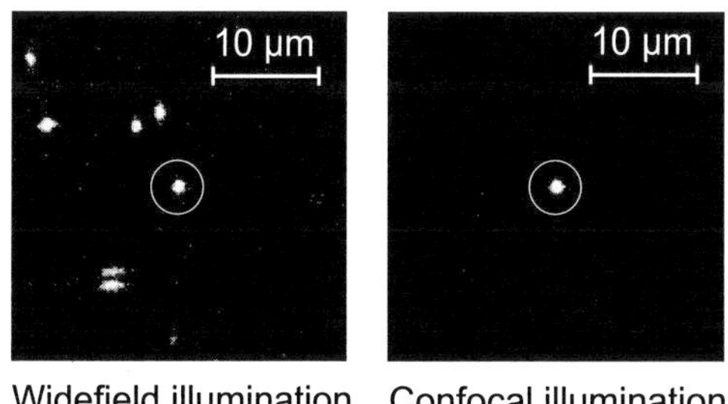

Figure 6.7: *A single nanocrystal on a microsphere is first chosen and then selectively excited by switching to the confocal mode of the microscope.*

coupler. As a first experiment, fluorescence from a single nanocrystal was collected with the microscope and sent to the spectrometer. The spectrum is plotted in figure 6.8. The width of the single nanocrystal spectrum is determined to be 13 nm with a

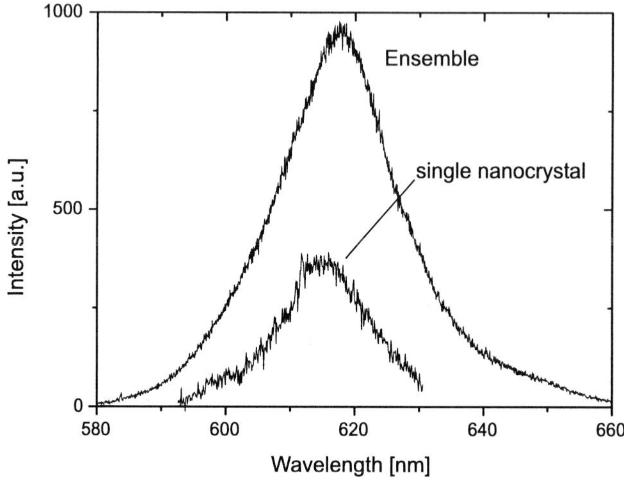

Figure 6.8: *Fluorescence spectra of a single nanocrystal and of an ensemble.*

Gaussian fit, while the ensemble width is for these nanocrystals $20\,nm$[4]. The measured values agree well with numbers reported in literature for single nanocrytsals at room temperature [SBPM02]. As further strong evidence for single nanocrystal fluorescence, blinking traces were recorded similar to those in section 3.2.4. Both experiments allow to identify single quantum emitters, but they cannot prove the coupling of the emitter to the cavity. However, the technique used in the case of the dye-doped bead and the ensemble of nanocrystals, namely the detection of a fluorescence spectrum via the prism coupler, is not applicable here. The reason is the weak signal arising from single nanocrystal fluorescence combined with a rather rapid photobleaching. With a typical observation time of only a few minutes, it is not possible to align the setup and then to distribute the few photons, extracted out of the microsphere by the prism coupler, on the detector array of the spectrometer CCD. From test experiments with a single $190\,nm$ bead it is known that the expected intensity is about 50 times less, compared to the signal strength obtained when using the microscope for direct detection. Therefore, a more sensitive and simpler method is required.

The strategy is to record the blinking trace of a nanocrystal on the microsphere with APD 1 of the confocal microscope and to use at the same time APD 2 to detect the photons coupled out of the whispering-gallery modes by the prism coupler (see figure 6.1). If there is a detectable signal behind the prism and if this signal is correlated with

[4]The spectral width of the ensemble depends strongly on the size distribution of the nanocrystals, when they are synthesized. This explains the deviation from the value already determined above.

the blinking trace recorded with the confocal microscope, then the photons detected with APD 2 were emitted by the single nanocrystal on the sphere. The result of such a measurement can be seen in figure 6.9. The signal of APD 1 is divided by a factor of

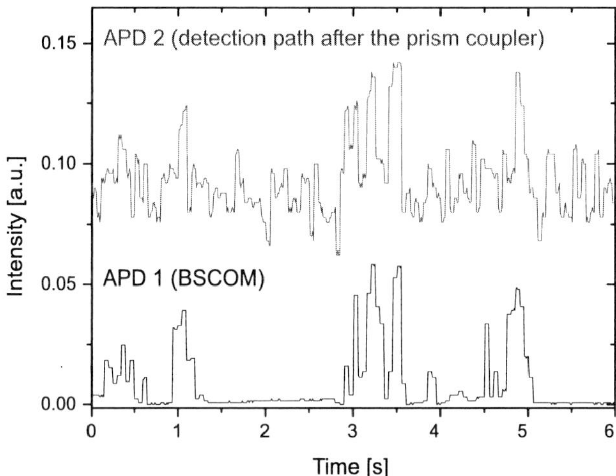

Figure 6.9: *Blinking behavior of a single nanocrystal observed with APD 1 of the beam scanning confocal microscope (BSCOM) and simultaneously observed via the prism outcoupler with APD 2.*

100. A strong correlation between the signals from APD 1 and APD 2 is present. This is a clear signature for the coupling of the nanocrystal to the microsphere. There is non-perfect correlation, because neither the two APDs nor the subsequent data acquisition was synchronized. Due to the finite integration of the system an "on period" of the nanocrystal could be summed up by one APD to one pixel, while the other APD divides the "on period" into two pixels. The correlation of the signals is obvious in figure 6.9. An improved measurement of the second order correlation function $g^{(2)}(\tau)$ [Lou97] with a time interval counter is, however, very difficult to perform due to the fast photobleaching of the nanocrystal.

6.2 A single bead attached to the end of a near-field probe

The previous section described experiments where single nanoparticles, even single nanocrystals, were successfully coupled to a microsphere resonator. Since the particles were randomly distributed on the sphere, they were only by chance at a position where a coupling to the sphere was observable. There was also, at least in principle, the possibility that other particles were excited from light scattered inside the sphere. In an ideal experiment one wants to deal with one and only one nanoemitter, which can be placed at will at a desired position.

This ideal situation can be achieved by attaching a single nanoparticle at the end of a near-field probe and using the scanning probe techniques introduced in the context of SNOM, described in section 3.3. The technique to attach a single nanoparticle to a near-field probe is described in detail in references [KRMS01, Kal02]. It is based on using an electrostatic or chemical bonding of the particle to the probe. This technique works for particle sizes of a few hundreds of nanometers down to at least $50\,nm$. Smaller sizes have not yet been attempted. Particles, e.g. large beads, with a size bigger than $2\,\mu m$ can simply be speared by the sharp fiber tip.

The same setup already described in the previous section can be used to perform experiments with a single dye-doped bead attached to a near-field probe, since the setup is flexible in the sense of the choice of the microscope. One can easily switch between the beam scanning confocal microscope and the SNOM. In Figure 6.10 the experimental setup is sketched, now with the SNOM inserted.

A microsphere was placed into the spectroscopy unit and coupled to the prism. The Q-factor was measured before and after the experiment to check that the coupling conditions did not change. As a second step, mode identification was performed. This could be done with a usual near-field probe or already with the fiber tip with the nanoparticle attached to it. After identifying the modes with $m \approx l$, the procedure described in the end of section 4.4 was used to optimize the coupling to the fundamental mode. This assured controlled and reproducible conditions for the experiment.

Before and after the experiment, the fiber tip with the bead was inspected with the confocal microscope in the wide field imaging mode to verify that the tip did not break and that the bead did not come off the tip during the experiment. The first picture in figure 6.11 is taken without filters, so that the fiber tip can be seen, the second with fluorescence filters, which block the excitation light. In the latter case, only the fluorescent dye-doped bead can be seen. The size of the beads used in the experiments was $500\,nm$. For precise information regarding the position of the bead on the fiber tip electron microscope images were taken.

In order to achieve good coupling conditions, the bead should be placed in the equatorial region of the microsphere. Therefore, the near-field probe with the bead was approached to shear-force distance and scanned vertically, while the topography signal

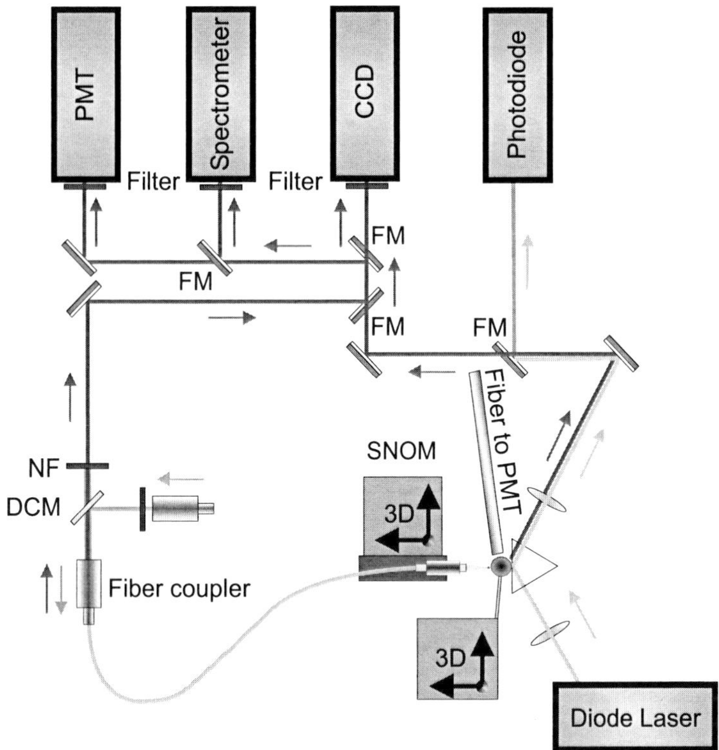

Figure 6.10: *Experimental setup for coupling a single nanoparticle attached to the end of a near-field probe to a high-Q silica microsphere. Light from a diode laser is coupled into the microsphere to control its coupling to the prism. The SNOM head allows to position the dye-doped bead with high precision on the microsphere's surface and to pump it via the optical fiber of the near-field probe. Fluorescence coupled to the resonator's modes can be extracted via the prism and sent to various detectors using flip mirrors (FM). Dichroic mirror (DCM), notch filter (NF)and photomultiplier tube (PMT).*

was recorded. The near-field probe was then aligned in an iterative process until the scan range was approximately symmetric around the equator. Excitation of the bead was simply done via the optical fiber of the near-field probe itself. This has the advantage, in comparison to a resonant pumping scheme via a whispering-gallery mode, that the pumping is independent of the mode structure and of the coupling conditions of the sphere to the prism. In case of resonant pumping a change of the Q-factor always

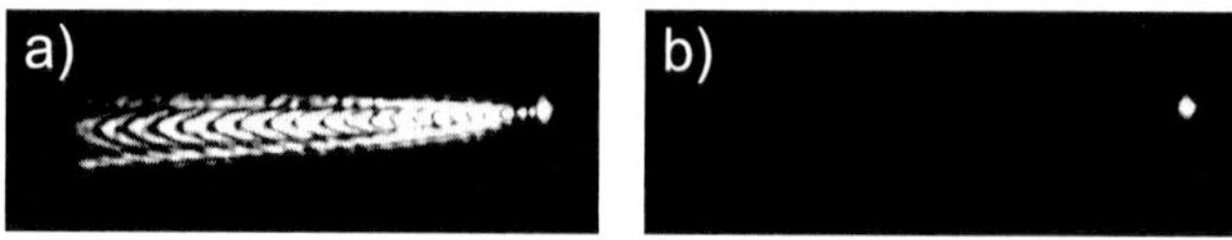

Figure 6.11: *a) Image of a fiber tip with attached bead taken without filters, such that light scattered by the tip is detected. b) Image of the same tip, but now taken with the notch filter and color glass filter inserted in the detection pathway, such that only fluorescence from the bead is detected.*

alters the coupling conditions and therefore the pump power. It turned out that an excitation power on the order of 20 μW compromises low photobleaching and appropriate signal strength at the same time; the power density is actually hard to determine exactly, since the tip does not perfectly guide the pump light to the very end in the case of pulled fiber tips. However, inspection with the confocal microscope has shown that the intensity of the fluorescence was approximately the same as it was in the case of experiments, when the beads were excited through the microscope.
As in the experiments with the confocal microscope, one can collect the fluorescence via two different paths. It can be sent either via the near-field probe to the various kinds of detectors or by using the prism as an the outcoupling port for fluorescence coupled to the microsphere. For the subsequent experiments only the latter was used. This arrangement allowed to control the coupling conditions of the bead to the microsphere by moving the fiber tip in all three spacial dimensions. The principle of the following experiments is sketched in figure 6.12. First investigations concerned the dependency

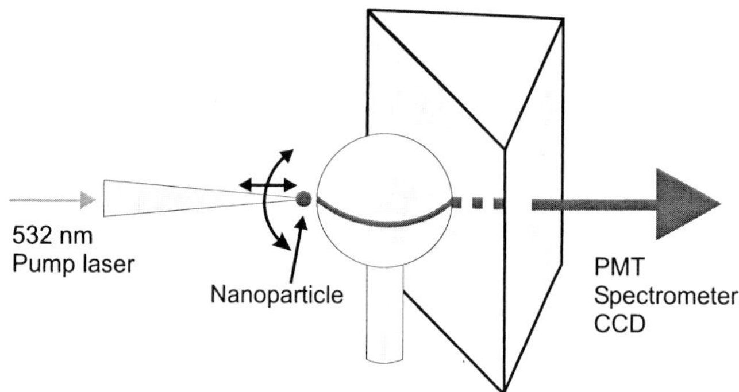

Figure 6.12: *Simplified sketch of the experimental setup.*

of the coupling on the bead's lateral position. The tip with the bead was scanned in the shear-force regime perpendicular to the sphere's equator (angle θ on the microsphere), while the PMT recorded the fluorescence intensity coupled out of the sphere by the prism coupler. Variation in the Q-factor by more than two orders of magnitude did not alter the results. The experiments were therefore performed with Q-factors between a few times 10^7 and 2×10^8. Figure 6.13 shows the obtained intensity distribution when the tip is scanned, in a) for a $86\,\mu m$ diameter sphere ($l \approx 652$ around $605\,nm$) and in b) for a $118\,\mu m$ diameter sphere ($l \approx 895$ around $605\,nm$). In order to fit the observed intensity distribution, the normalized intensity distribution of each of the first 10 whispering-gallery modes on the sphere's surface were used as a basis set (see equation (2.49)). The first mode is the fundamental mode and the 10th mode is the mode with 10 lobes perpendicular to the sphere's equator. According to equation (2.49), the following fit function was used

$$I_{l,m}(\theta) = a_0 + \sum_{i=1}^{10} a_i [H_{i-1}(l^{\frac{1}{2}} \cos\theta) \times \sin^{l+1-i}(\theta)]^2, \tag{6.1}$$

where m was substituted by $l+1-i$ and a_0 sets an offset. The coefficients a_i weight the contributions of the individual polynomials (modes). The agreement of the fit with the experimental data is remarkable; even the ripple structure of the fit can be identified in the experimental data.

The intensity distribution of the first 5 modes, weighted according to the obtained a_i, is plotted in figure 6.13 a) to illustrate the contribution of the individual modes. One can clearly see the monotonic decrease of the absolute heights of the polynomials, since the values of the fit parameters are decreasing. The values of all ten coefficients for figure 6.13 a) are shown in figure 6.14. The contribution of each mode to the fit is a combination of the coupling of the bead to a certain whispering-gallery mode and the selectivity of the prism coupler as the output port. Since two consecutive modes do not differ too much from each other, a reasonable assumption is that the prism coupler dominates that process. From the quality of the fit and the shear-force signal, which was recorded simultaneously with the outcoupled intensity, one sees again how good the equator of the sphere is aligned with respect to the prism. The intensity distributions shown in figure 6.13 are symmetric around the equator ($\theta = 90°$) and show a single central peak. In test experiments where the sphere was tilted by 12° out of the optimal plane, the intensity distribution split into a double peak (one at $\theta = 90°$), as one expects when the fundamental mode does not couple anymore to the prism and modes with $l - |m| + 1 \geq 10$ provide the main contribution to the detected signal.

The experiments discussed above clearly demonstrate a strong position dependency of the coupling of a single nanoparticle on the microsphere-prism system. Since they also show at what position the best coupling can be achieved, they open the way to controlled experiments where the bead is placed with almost nanometer accuracy at a location where the coupling is maximal. In particular, the nanoparticle can be moved into a region, where it couples strongest to the fundamental mode, but not to any modes with an even number of lobes.

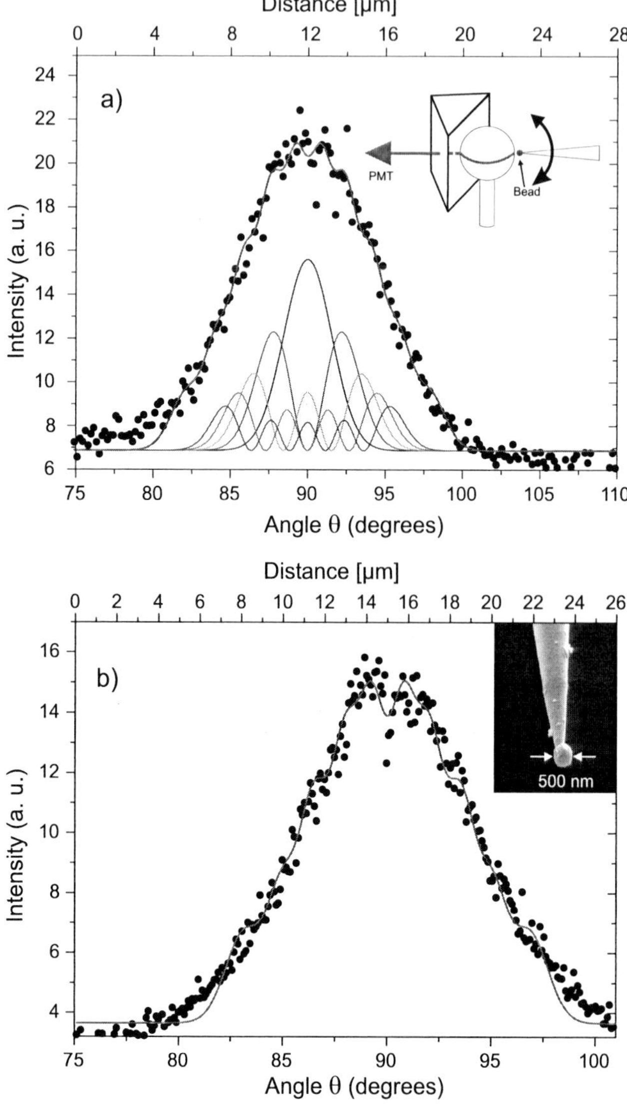

Figure 6.13: *Intensity of the fluorescence coupled out of the microsphere by the prism versus the bead's position shown for two different microspheres. The red line is a fit with the first ten modes, starting with the fundamental down to* $l - |m| + 1 = 10$. *In a), the first 5 weighted modes are plotted, to illustrate their contribution to the fit. The inset in b) shows an electron microscope image of the bead attached to the near-field probe, which was used in the experiment.*

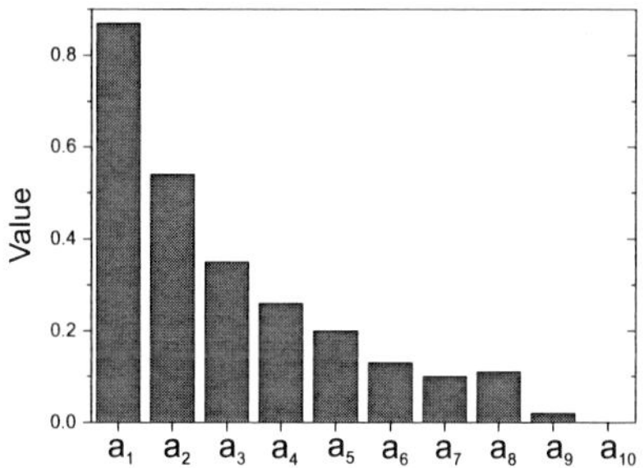

Figure 6.14: *Diagram of the coefficients a_i.*

Additionally, pulling the tip forwards and backwards offers a way of changing the coupling to a particular mode without increasing the coupling to another one. Figure 6.15 clearly demonstrates how the intensity of light inside the whispering-gallery modes decreases when the bead is moved away from the sphere's surface. The vertical line separates the shear-force region on the left side from the region where the tip moves away from the microsphere's surface. Within the shear-force region, the active feedback loop keeps the distance constant at a value of about $10\,nm$. The rapid decay to the background level within $200\,nm$ indicates the evanescent nature of the coupling process.

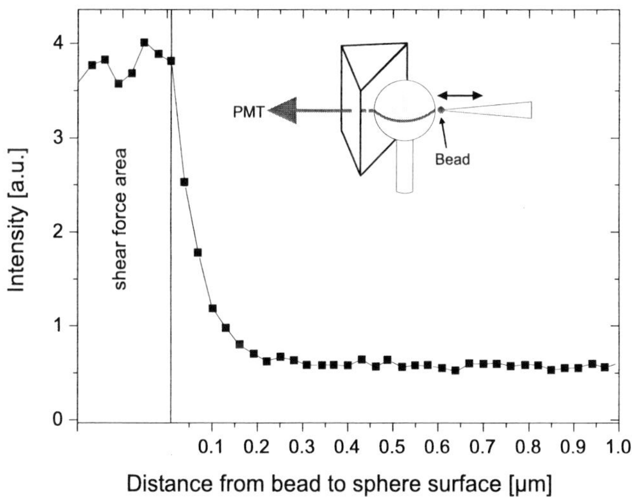

Figure 6.15: *The fiber tip with the bead attached is pulled backwards, while the PMT detects the fluorescence coupled to the microsphere-prism system. After only $200\,nm$ the signal has background level.*

Finally, it is possible to perform spectroscopy of a single bead coupled to a high-Q microsphere resonator under optimized coupling conditions. Figure 6.16 shows a high resolution spectrum around $602\,nm$, close to emission peak of the used beads. The

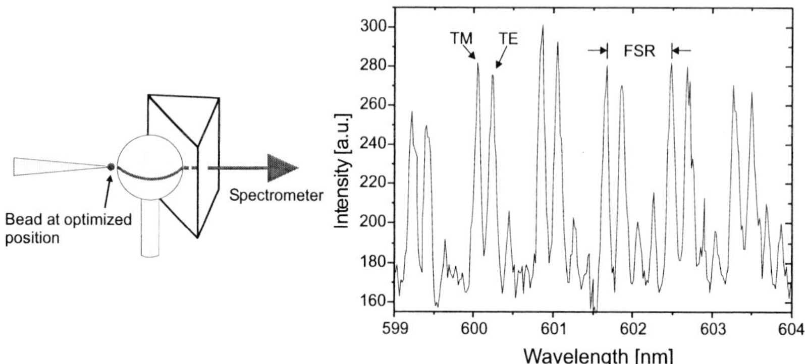

Figure 6.16: *Controlled coupling of a single $500\,nm$ bead to a high-Q microsphere resonator. The graph shows a high resolution spectrum of the emission coupled out of the microsphere resonator with the use of the prism coupler.*

spectrum is taken via the prism outcoupling port. Consistent with the bead and the nanocrystals addressed with the confocal microscope the spectrum shows a periodic structure. All the features, previously discussed in this context, are present. The FFT lead to an FSR of $704\,GHz$, which is in good agreement with the theoretically expected FSR of $(726 \pm 65)\,GHz$ for the sphere, which had a diameter of $90\,\mu m$. This corresponds to a modespacing of $0.85\,nm$ at a wavelength of $603\,nm$. TE and TM modes were identified again as well as the fainter peaks, which are attributed to modes with $n = 2$.

To summarize, in this chapter coupling of nanoparticles to high-Q whispering-gallery modes with two different methods was described. Using a beam scanning confocal microscope allowed even to demonstrate the coupling of a single nanocrystal to whispering-gallery modes. In the second approach, techniques well known from scanning near-field optical microscopy gave ultimate control over the coupling conditions of a $500\,nm$ dye-doped bead. The results clearly demonstrate that experiments where a single quantum emitter is coupled to a high-Q microsphere under controlled conditions are feasible.

Chapter 7

Towards coupling of two nanoparticles via whispering-gallery modes

The ability to couple single nanoparticles to a high-Q microsphere can be regarded as a first important step towards more sophisticated experiments, where the unique properties of these resonators are used. Lasing of a nanoscopic amount of active material is just one example. One can also imagine the coupling of two nanoparticles (e.g. two atoms or molecules) via the whispering-gallery modes. Microspheres seem to be ideally suited for such kind of experiments due to their long photon storage time and the small mode volume, so that only a single photon builds up a significant field [CLSB+93].
In the case of coupling two particles, a prerequisite is that the first particle (hereafter called donator) emits at a wavelength where the second particle (hereafter called acceptor) has a strong absorption cross section. Additionally, the acceptor should have its emission maximum shifted with respect to the donator, such that both particles can easily be spectrally discriminated. In the following chapter first experiments are described, which give strong evidence for such a coupling.
Two strategies are pursued. In the first approach, the beam scanning confocal microscope pumps a donator particle, while the same microscope objective collects light from the area where an acceptor is located.
The second approach uses the very elegant method of upconversion in an erbium-doped fiber tip to excite beads on the microsphere's surface. The big advantage in this case is that the pump light for the donator cannot excite the acceptor and the fiber tip can be positioned as part of the SNOM at any desired position on the sphere's surface.

7.1 Two beads coupled via whispering-gallery modes

In the previous chapter it was shown how single nanoparticles can be coupled to the whispering-gallery modes by means of confocal microscopy. The same setup can be used (see figure 6.1) to pump a dye-doped bead on the surface of the sphere, whose fluorescence couples to the whispering-gallery modes and excites an acceptor bead on the microsphere. The requirements which the beads need to fulfill are discussed above. As donator a 190 nm bead (Molecular Probes, Red Fluorescent) was chosen, which has its absorption maximum at 580 nm and its emission peak around 610 nm. Excitation with a 532 nm pump laser is not a problem due to the broad absorption band of the beads. As acceptors, 200 nm beads (Molecular Probes, Crimson) were chosen with the absorption peak around 625 nm and the emission maximum at 645 nm. The absorbance of the acceptor is about 70 % of the maximum at the emission peak of the donator. Also, the 532 nm pump laser can still excite these beads, which is important for easy identification on the sphere.

The experiment started with a dipping process, where the microsphere was dipped for a while in a solution of methanol, where both beads were suspended in the desired concentration. Then the microsphere was installed into the setup. Coupling and alignment was performed in the same way as described in section 6.1. Q-factors were always at least 10^7. In figure 7.1, a simplified sketch illustrates the experimental setup.

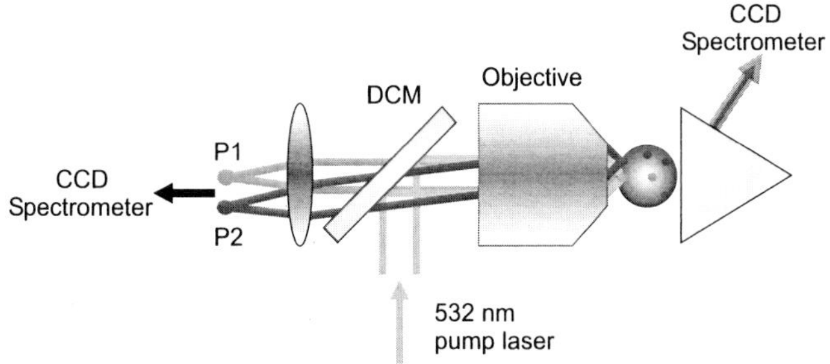

Figure 7.1: *Sketch of the experimental setup to detect the coupling of two beads via whispering-gallery modes. P1 and P2 denote the images of a donator and an acceptor bead on the CCD camera or the spectrometer. DCM, dichroic mirror.*

First, a single donator bead and an acceptor, both close to the microsphere's equatorial region, were chosen. Then, only the donator was excited by switching to the confocal mode of the microscope. Fluorescence signals from the entire field of view of the microscope were recorded with the CCD camera.

Figure 7.2 a) shows the result of the experiment. Although only the donator (D) is pumped, there is significant signal, where the acceptor (A) was identified before. In order to exclude the possibility that the acceptor is pumped by direct or indirect excitation with the green laser, the focus was moved to the close vicinity of the acceptor. As seen in figure 7.2 b), fluorescence is neither observable from the donator nor from the acceptor. One can thus deduce that there is a relation between the excitation of the donator and the signal at the acceptor's position. Spectral information can reveal the origin of the observed signals. The donator is pumped by the focussed laser, while a spectrum is recorded only from light collected at the acceptor's position. This was done by imaging the acceptor (P2 in figure 7.1) on the spectrometer slit (width $100\,\mu m$) and using the spectrometer software to choose a small region of interest around the acceptor, such that only photons, which have their origin at the acceptor's position, contributed to the recorded spectrum. The result is plotted in figure 7.3, where an adjacent averaging was performed to smooth the signal. The upper curve shows the spectrum at the acceptor's position, while the lower one (dashed) shows a spectrum taken at the same position, but after having bleached the acceptor. As a first result, one can clearly see that the main contribution to the signal observed in figure 7.2 a) arises from photons emitted by the donator (peak emission around $610\,nm$) into the whispering-gallery modes. These photons are scattered out of sphere due to the presence of the acceptor bead. This is not surprising, since the same effect was used before routinely for mode identification, where an optical fiber with a sharp tip at the end scattered light out of the mode. However, conversely, this can also be seen as a proof that the bead emits a significant number of photons into the whispering-gallery modes.

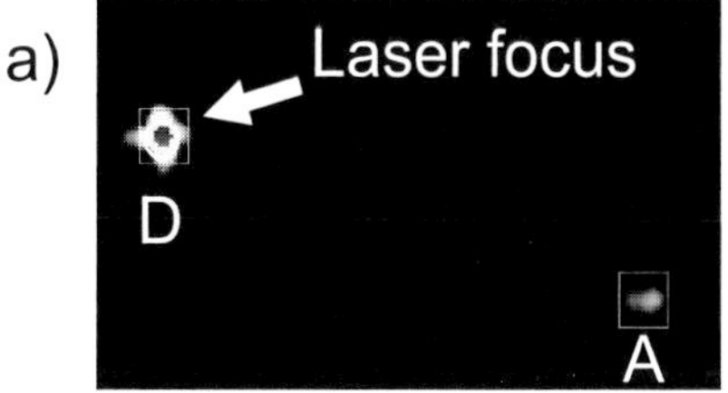

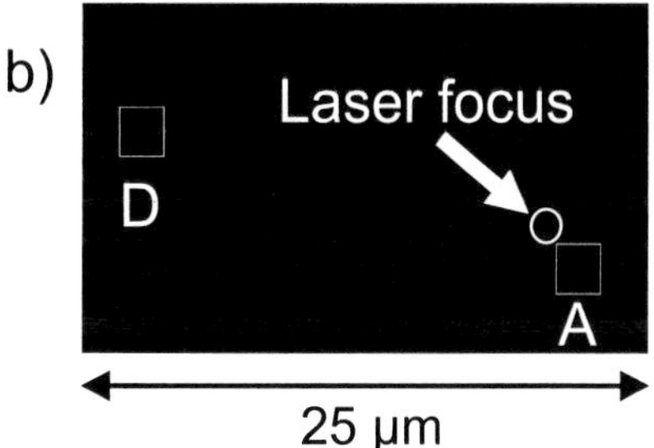

Figure 7.2: *a) The donator (D) is excited by the focussed pump laser. b) The focus is moved close to the acceptor (A).*

Furthermore, there is a difference between both curves in the spectral region where the acceptor emits. This is a hint that the acceptor was indeed pumped by the donator before it was bleached (upper curve). Subtraction of the lower curve from the upper one corroborates this conjecture. The result is plotted as the dashed curve in figure

7.4.

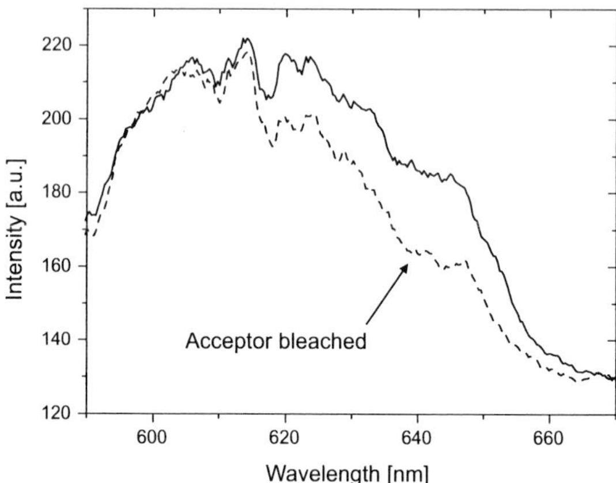

Figure 7.3: *The upper curve shows a spectrum of light taken at the acceptor's position, when only the donator was excited. The dashed curve is obtained under the same pumping condition after the acceptor has been bleached.*

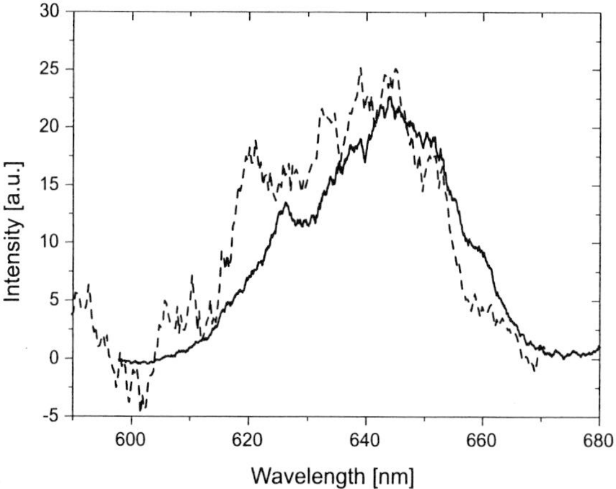

Figure 7.4: *Dashed curve: difference between the two spectra plotted in figure 7.3. Solid curve: emission spectrum of the acceptor.*

For the purpose of comparison, the spectrum of a confocally excited acceptor is shown in the same graph (solid line). Both curves are similar, but with a shift in the maximum intensity. This can happen when the spectrometer is not perfectly aligned. Test experiments confirmed that the emission peak may shift by several nanometers when the coupling to the spectrometer is changed slightly. Therefore, there is strong evidence that the photons were emitted indeed by the acceptor. However, one has to exclude the possibility that pump photons were scattered by the donator bead into the whispering-gallery modes, while it was excited by the confocal microscope and that these photons from the pump excited the acceptor directly. Although this is in principle possible, similar to an inverted SNOM-process, where light is scattered into the modes, this is not very likely for the following reasons: the linewidth of the pump laser is $5\,MHz$ and the linewidth of a whispering-gallery mode with a Q-factor of only 10^7 is $40\,MHz$, but the mode spacing between two consecutive modes differing in m is on the order of $4\,GHz$. By comparing the width of the resonance with the spacing between two consecutive modes, one can deduce that the probability that the pump laser frequency overlaps with a whispering-gallery mode is $1 : 100$. But even if this unlikely event happens, one could change the set point of the sphere's temperature stabilization, such that the whispering-gallery mode is tuned out of resonance with the pump laser. For smaller particles like nanocrystals or molecules, scattering of pump light is negligible, since the total scattering cross section scales, according to section 5.3, with the radius to the sixth power and is thus very small.

In summary, these experiments give strong evidence for the coupling of two nanoparticles via whispering-gallery modes. Although the obtained data is already convincing, ongoing experiments need to deliver an ultimate proof. In one test experiment, one could for example bleach the donator and observe if the acceptor still emits photons, when the confocal spot is placed on the donator. The result would show if the pump laser can couple to whispering-gallery modes due to scattering on the donator.

Another test is an experiment, where the donator bead is attached to a near-field probe, giving ultimate control over the coupling conditions. Or one could even use a second near-field probe, to which the acceptor is attached. This would assure that only two particles would interact and in this case in a very controlled manner.

The latter experiment is ideal in order to study the coupling of two quantum emitters in great detail. However, a main prerequisite for such experiments is a high photostability of the two quantum emitters at room temperature.

7.2 Coupling an erbium doped fiber tip to a single bead

The previous section has demonstrated a method to observe the coupling of two nanoparticles via whispering-gallery modes. However, the question arose if, at least in principle, the pump light for the donator could also pump the acceptor. In fact, a direct pumping by the laser could not be completely excluded in the experiments performed so far, although it is not likely. Such problems could be completely avoided if the pump laser wavelength could not excite the acceptor beads. A rather elegant way is offered by erbium ions, which could be pumped by a non-linear upconversion process [Sch96], at about $800\,nm$. In this case the erbium ions, having a strong emission around $550\,nm$, are perfectly suited to excite the beads previously used in this work, which emitted at $610\,nm$. The excitation light does not excite the beads in this case, assuming no two photon absorption.

The other major difference, besides using upconversion for pumping the donator, compared to the previously described experiments is that the donator is now part of a near-field probe. Erbium doped optical fibers are commercially available. If one etches tips out of these fibers, then only a small amount of ions would couple via the evanescent field to the microsphere. With a typical end diameter of about $100\,nm$ and an evanescent decay length of the whispering-gallery modes of about $200\,nm$ the end of the fiber tip can be regarded as a single nanoparticle with a diameter smaller than $200\,nm$.

In the experiment, two different fibers were tested. The first one was a single mode silica fiber doped with $5500\,ppm$ Er^{3+} (OFS FITEL). Short pieces of these fiber were spliced to a non-doped single mode fiber and etched with HF in the standard way. The second kind of fiber was a multimode ZBLAN fiber doped with $50000\,ppm$ Er^{3+} (Le Verre Fluorè). The core diameter was $60\,\mu m$ and the cladding diameter $120\,\mu m$. ZBLAN glass based on zirconium fluoride as the host material offers a more efficient upconversion, due to its lower maximum phonon energy (ZBLAN is a so-called low-phonon-energy glass [Cas01].). However, its handling is rather complicated compared to silica [DPNM+96, Göt98]. It is neither possible to etch those fibers with HF, nor can one splice them easily to a silica fiber. The way to etch ZBLAN tips is described in reference [Kra02]. The problem of splicing is avoided by putting a short piece of this fiber already cleaved and etched in contact with a silica single mode fiber and gluing both inside a micropipette. Since the ZBLAN fiber has a rather large core, sufficient light from the single mode silica fiber couples.

The experimental setup is based on a mixture of the configurations already shown in figure 6.1 and figure 6.10. To demonstrate the coupling of the erbium tip to the sphere only the SNOM is needed, but to observe the coupling of the erbium tip to a bead, both microscopes need to be used simultaneously. In figure 7.5, the simplified setup is depicted.

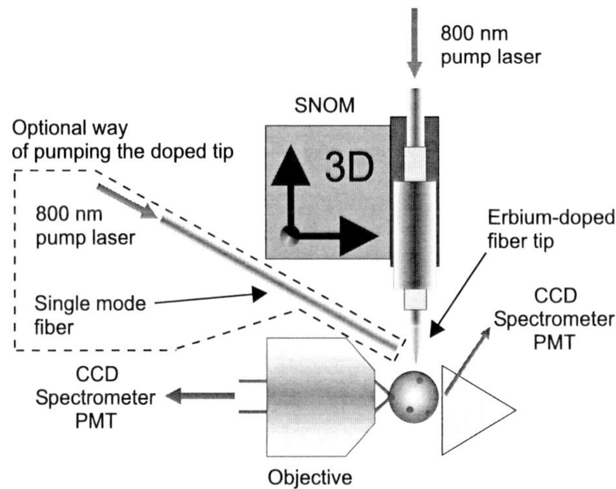

Figure 7.5: *Sketch of the experimental setup to couple an "erbium nanoparticle" to a single dye-doped bead. The erbium tip is part of a near-field probe, which can be placed in close vicinity of the sphere by using shear-force. A side pumping scheme (dotted box) offers an alternative way of pumping the ions at the very end of the tip. Fluorescence of the beads on the sphere's surface can be collected with the beam scanning confocal microscope and sent to various detectors.*

In a first experiment, it was demonstrated that the tip couples to the microsphere. This was done in the same manner as in section 6.2, where a bead was attached to the end of a near-field probe. Here the erbium tip was pumped with a Ti:Sapphire laser (Coherent Mira) operating at a wavelength of $800\,nm$. Both CW as well as pulsed operation could be used to excite the ions. After the microsphere was coupled to the prism, the fiber tip was brought to shear-force distance, excited with the pump laser and a spectrum of the light coupled to the microsphere was recorded via the prism outcoupler. The setup offers two ways of pumping the tips, either directly via the optical fiber of the near-field probe or with a single mode fiber which was brought to a distance of about $70\,\mu m$. The advantage of the latter method is that only the last $20\,\mu m$ of the fiber are pumped. In this case there are far less ions excited which do not couple to the sphere. The result is a reduction of green background light. Furthermore, the achievable power density at the end of the tip is much higher, since the pump light is not guided through a certain distance in the doped fiber, where it is absorbed by ions and thus attenuated, until it reaches the end of the fiber. This side pumping is especially necessary in the case of the ZBLAN fiber, because of the big core size and

the minimal length of the erbium doped part of $10\,mm$[1], which is necessary to be able to glue it into the micropipette and to have the resonance frequency of the tip in a region useful for shear force. The drawback of this method is that the single mode fiber, which is used for the side pumping, needs to be moved synchronized with the SNOM, when the tip is moved.
The proof of the coupling of erbium ions to the sphere can be seen in figure 7.6. The upconversion signal around $550\,nm$ clearly shows the periodic structure as was previously discussed, when nanoparticles were coupled to high-Q whispering-gallery modes. The reason that the spectrum does not show the previously observed $100\,\%$ modulation

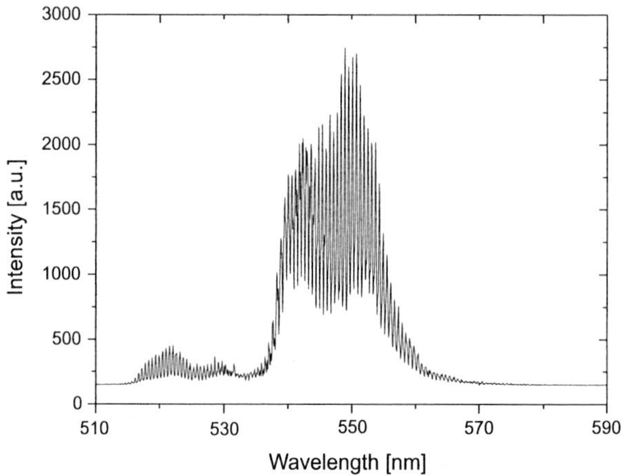

Figure 7.6: *Upconversion spectrum around $550\,nm$ of an erbium doped ZBLAN fiber tip coupled to a high-Q microsphere resonator detected via the prism coupler.*

is the resolution of the spectrometer. In this measurement, where a wide spectral range needed to be covered, it was only $0.2\,nm$ due to the grating used ($600\,grooves/mm$) and a spectrometer slitwidth of $100\,\mu m$. The observed spacing of $0.6\,nm$ agrees well with the value one expects for a $110\,\mu m$ sphere. The TE/TM splitting cannot be observed due to the resolution of the spectrometer.
One can deduce from this measurement that photons emitted by the ions couple to the whispering-gallery modes and can thus be used to excite the molecules in a bead on the surface of the sphere.
In the next step, a microsphere coated with beads was inserted into the setup. First, the

[1] In the case of the silica fiber, the doped part could be as short as $500\,\mu m$, restricted by the accuracy of the etching process.

beads were identified on the sphere's surface, by using the confocal microscope in the wide field imaging mode. Then, the green pump laser of the microscope was switched off and a doped tip, pumped via the associated fiber, was held at a distance of $6\,\mu m$ to the microsphere. An image of the sphere was recorded where a bead was identified (inset in figure 7.7 a)). Afterwards the tip was brought to shear-force distance and the

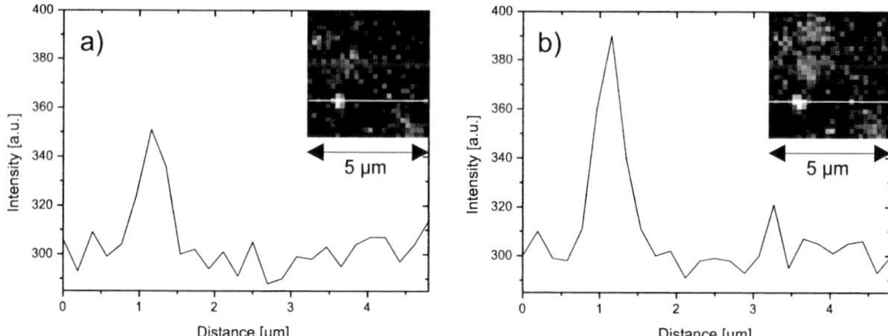

Figure 7.7: *Cross sections through a bead an a microsphere pumped by an erbium doped tip. a) The tip has a distance of $6\,\mu m$ to the sphere's surface. b) The tip is at shear-force distance. The insets show the images where the cross sections were taken from.*

procedure was repeated (inset in figure 7.7 b)). The bead is clearly visible in the image. A comparison of the cross sections shows an enlargement of the intensity, when the tip is at shear-force distance. But the bead is also excited by a direct illumination with green photons from the fiber tip, when it is $6\,\mu m$ away. Therefore, one cannot directly deduce whether the enhanced signal in figure 7.7 b) arises from a coupling of green photons to the whispering-gallery modes or simply from an increased power density, because of the shorter distance between tip and bead. A test experiment would be to measure an approach curve as it was already recorded for a single bead attached to the end of a near-field probe (see figure 6.15). This would definitely show which effect increases the signal. Another improvement would be the use of the side pumping scheme, because then as previously mentioned only about $20\,\mu m$ of the doped tip would emit photons and thus reduce the number of photons out of undesired regions by orders of magnitudes.

In summary, an experiment was discussed where an upconversion process is used to couple erbium ions via whispering-gallery modes to dye-doped beads. The coupling of the ions to a microsphere could be demonstrated. In a first measurement the fluorescence enhancement of a single bead was observed when the doped fiber tip was brought to shear-force distance to the sphere's surface. However, further measurements need to prove whether this is indeed an effect arising due to photons in whispering-gallery modes.

Chapter 8

Outlook

The controlled coupling of single nanoparticles to a high-Q microsphere resonator has been demonstrated for the first time in this work. Controlling the sphere's modes and their coupling to the prism was an important prerequisite, which was achieved by using a scanning near-field optical microscope for mode identification. It was also shown that such a near-field probe could act as a perfect nanohandle, since it does not degrade the Q-factor, even when positioned in the intensity maximum of the fundamental mode. Therefore, the approach where a single nanoparticle was attached to the end of a near-field probe appears promising, and opens the way to various new experiments, where an emitter is placed at will with nanometer precision in the evanescent field of the microsphere. Moreover, such functionalized nanoprobes will be very useful in any experiment where the position of a single nanoparticle has to be controlled and/or modified with ultimate precision.

A nano laser and cavity-enhanced effects

The realization of a laser with a nanoscopic amount of active material appears to be straightforward. One could substitute the beads used so far, which are usually used in biological applications, by rhodamine 6G doped beads. Such a bead with a size of the order of $100\,nm$ would be attached to the end of a fiber tip and would then be coupled to the fundamental mode of the microsphere, while the Q-factor of the cavity would still be 10^8 or higher. In order to have a directed beam, the prism would act as the outcoupler for the laser emission. Pumping of the bead could be performed with a pulsed frequency doubled Nd:YAG laser (pulse length $\approx 5\,ns$), either via the near-field probe itself or by focusing the free beam on the end of the near-field probe. This will depend on the pump power needed to achieve population inversion, since ns pulses can easily create power densities high enough to destroy the end facette of the optical fiber. This system would be perfectly suited to study the lasing properties of a nano laser, since one can vary, for example, the Q-factor by changing the sphere-prism gap. Moreover, the nanomanipulation with the near-field probe allows the mode volume of

the lasing mode to change simply by moving the bead to another position, so that it couples strongest to more extended modes.
Also, the investigation of cavity enhanced nonlinear effects [LC94, USH02] is possible with this setup. It has already been shown that rare-earth doped fiber tips couple to the microsphere. Probably, a resonant pumping scheme via whispering-gallery modes is favorable to enhance the pump intensity due to the high-Q of the cavity. Upconversion lasing with a doped fiber tip might also be possible. This would then be the smallest upconversion laser ever built.
Other kinds of experiments which were already started in this work, concern the coupling of two nanoparticles via the whispering-gallery modes. Also here, the approach with the near-field probe would give more control over the experimental parameters. Such experiments can be seen as a first step towards the coupling of two quantum dots to a single high-Q cavity mode. Having the dots at the end of two near-field probes could allow the realization of quantum information processing schemes, such as those proposed in reference [IAB+99].

A cryogenic single quantum emitter laser and cavity QED

The ultimate goal of one of the experiments started in this thesis, namely the realization of a single quantum emitter laser, cannot be obtained at room temperature[1], neither with molecules nor with quantum dots which are the emitters currently considered. The main reasons are the broad emission linewidth and photobleaching of molecules and the extremely low luminescence quantum efficiency of epitaxily grown semiconductor quantum dots at room temperature. Therefore, in order to move this experiment forward, a low temperature setup needs to be constructed. Then the emitter's linewidth will decrease significantly and could therefore couple only to one mode of the cavity. Also, photostability is significantly improved for molecules, and quantum dots are, in any case, stable at cryogenic temperatures. An experiment where molecules in a host crystal were attached to the end of a near-field probe and used in a cryogenic setup was performed recently [MHMS00]. In addition, some promising attempts to grow quantum dots on tips in a controlled way were also already obtained.
All the essential parts to realize a cryogenic setup were developed and tested at ambient conditions in the context of this work. For example, the sensitivity needed to detect the emission of a single quantum emitter laser was demonstrated by the coupling of a single nanocrystal to the microphere resonator. Controlling all the degrees of freedom needed to couple the fundamental mode to the prism remains challenging at low temperatures. Although one of the most important tools for a low temperature setup, namely a nano positioning unit, based on a slip-stick mechanism, has already been constructed and tested.

[1] A one-atom laser was recently successfully demonstrated, but still without real control on the atom's position. The atom could be trapped for only about $100\,ms$ in the cavity [MBB+03].

The ability to attach a single quantum emitter to the end of a near-field probe also opens the way to new experiments in the field of cavity QED with microsphere resonators. Lifetime changes due to the presence of the high-Q microcavity are expected. Even the regime of strong coupling could be investigated. A 20 μm sphere with a Q-factor of only 8×10^6 would already compete favorably with the current state of the art in Fabry-Perot cavity QED [BK03]. The assumed Q-factor is two orders of magnitude smaller than the theoretical maximum calculated in this reference. A noticeable advantage of microspheres is that they are easy and cheap to produce compared to high-end Fabry-Perot cavities. But the main advantage is the ability to control the Q-factor and therefore the cavity decay time, by changing the sphere prism gap.

Microspheres as compact optical sensors

So far, only the use of silica microspheres in fundamental science was discussed. When the microsphere spectroscopy experiment was developed, strong efforts were made to stabilize the sphere against thermal and mechanical drifts. Such effects, like a change in the sphere prism gap or a change in temperature, can be measured quite accurately by performing spectroscopy on the narrow resonances of the microsphere resonator. In this way a high-Q silica microspheres represent compact sensors, where the extremely high-Q results directly in high sensitivity. The idea would then be to establish the particular sensing task via measurement of whispering-gallery mode perturbations, e.g. shifts or broadening of high-Q resonances.

The use of a microsphere as a temperature sensor is straightforward. One simply has to measure the frequency shift of a resonance. Another kind of sensor can be based on variations in the sphere prism gap, which reflects in the Q-factor as well as in the resonance's amplitude. By monitoring one of these quantities, one can built a accelerometer or vibration sensor [TLL03].

Trace-gas detection is another example. Due to the small size and the high-Q of the microsphere, there is only a small volume of gas needed for analysis. A similar application would be a chemical/biological agent sensor, where the microsphere is surrounded by a liquid. Q-spoiling by the liquid is not necessarily a problem [CMA$^+$01]. One can also imagine functionalizing the sphere surface, such that only a particular kind of molecule or DNA-fragment binds to the surface of the sphere and changes the spectral properties of the resonator. This work has demonstrated in particular that attaching nanoparticles does not necessarily degrade the Q-factor, so even after functionalizing the surface of the microsphere sensors, sensitivity could still be very high.

Bibliography

[AS72] M. Abramowitz and I.A. Stegun. *Handbook of mathematical functions.* Dover Publications Inc., 1972.

[BBA+92] K. Brunner, U. Bockelmann, G. Abstreiter, M. Walther, G. Böhm, G. Tränkle, and G. Weimann. Photoluminescence from a single GaAs/AlGaAs quantum dot. *Physical Review Letters*, 69(22):3216–3219, 1992.

[BBHS91] M. G. Boshier, D. Berkeland, E. A. Hinds, and V. Sandoghdar. External-cavity frequency-stabilization of visible and infrared semiconductor lasers for high resolution spectroscopy. *Optics Communications*, 85(4):355–359, 1991.

[BC88] P. W. Barber and R. K. Chang. *Optical Effects Associated With Small Particles.* World Scientific, Singapore, 1988.

[BFW92] E. Betzig, L. Finn, and L. S. Weiner. Combined shear-force and near-field scanning optical microscopy. *Applied Physics Letters*, 60(20):2484–2486, 1992.

[BGI89] V. B. Braginsky, M. L. Gorodetsky, and V. S. Ilchenko. Quality-factor and nonlinear properties of optical whispering-gallery modes. *Physics Letters A*, 137(7.8):393–397, 1989.

[BH90] P. W. Barber and S. C. Hill. *Light Scattering by Particles: Computational Methods.* World Scientific, Singapore, 1990.

[BH98] C. F. Bohren and D. R. Huffman. *Absorption and Scattering of Light by Small particles.* John Wiley & Sons, New York, 1998.

[BHM96] J. Barenz, O. Hollricher, and O. Marti. An easy-to-use non-optical shear-force distance control for near-field optical microscopes. *Review of Scientific Instruments*, 67(5):1912–1916, 1996.

[BK03] J. R. Buck and H. J. Kimble. Optimal sizes of dielectric microspheres for cavity QED with strong coupling. *Physical Review A*, 67:033806–(1–11), 2003.

[BMH99] R. Brunner, O. Marti, and O. Hollrichter. Influence of environmental conditions on shear-force distance control in near-field optical microscopy. *Journal of Applied Physics*, 86(12):7100–7106, 1999.

[BS91] I. N. Bronstein and K. A. Semendjajew. *Taschenbuch der Mathematik*. Teubner Verlagsgesellschaft, Stuttgart, 25th edition, 1991.

[BT92] E. Betzig and J. K. Trautman. Near-field optics: microscopy, spectroscopy and surface modification beyond the diffration limit. *Science*, 257:189–195, 1992.

[BTH+91] E. Betzig, J. K. Trautmann, T. D. Harris, J. S. Weiner, and R. L. Kostelak. Breaking the diffraction barrier: optical microscopy on a nanometric scale. *Science*, 251:1468–1470, 1991.

[BY99] O. Benson and Y. Yamamoto. Master-equation model of a single-quantum-dot microsphere laser. *Physical Review A*, 59(6):4756–4763, 1999.

[Cas01] R. Caspary. *Applied Rare-Earth Spectroscopy for Fiber Laser Optimization*. Dissertation, Universität Braunschweig, 2001.

[CBB94] D. Courjon, C. Bainier, and F. Baida. Seeing inside a fabry-perot resonator by means of a scanning tunneling optical microscope. *Optics Communications*, 110:7–12, 1994.

[CC96] R. K. Chang and A. J. Campillo. *Optical Processes in Microcavities*. World Scientific, Singapore, 1996.

[CK96] T. Corle and G. Kino. *Confocal Scanning Optical Microscopy and Related Imaging Systems*. Academic Press, San Diego, 1996.

[CLSB+93] L. Collot, V. Lefèvre-Seguin, M. Brune, J. M. Raimond, and S. Haroche. Very high-q whispering-gallery mode resonances observed on fused silica microspheres. *Europhysics letters*, 23(5):327–334, 1993.

[CMA+01] Y.-S. Choi, H. J. Moon, K. An, S.-B. Lee, J.-H. Lee, and J.-S. Chang. Ultrahigh-Q microsphere dye laser based on evanescent-wave coupling. *Journal of the Korean Physical Society*, 39(5):928–931, 2001.

[CMD72] C. K. Carniglia, L. Mandel, and K. H. Drexhage. Absoption and emission of evanescent photons. *Journal of the Optical Society of America*, 62(4):479–486, 1972.

[Col94] L. Collot. *Etude théorique et expérimentale des résonances de galerie de microsphère de silice: pièges à photons pour des expériences délectrodynamique en cavité*. PhD Thesis, Laboratoir Kastler Brossel de l'École Normale Supérieur, Paris, 1994.

[CPV00] M. Cai, O. Painter, and K. J. Vahala. Observation of critical coupling in a fiber taper to a silica-microsphere whispering-gallery mode system. *Physical Review Letters*, 85(1):74–77, 2000.

[Dem99] S. Demmerer. *Aufbau eines High-Q Whispering-Gallery Resonators für einen Nanolaser*. Diplomarbeit, Universität Konstanz, 1999.

[DHK+83] R. W. P. Drewer, J. L. Hall, F. V. Kowalski, J. Hough, G. M. Ford, A. J. Munley, and H. Ward. Laser phase and frequency stabilization using an optical resonator. *Applied Physics B*, 31:97–105, 1983.

[DKL+95] N. Dubreuil, J. C. Knight, D. K. Leventhal, V. Sandoghdar, J. Hare, and V. Lefèvre. Eroded monomode optical fiber for whispering-gallery mode excitation in fused-silica microspheres. *Optics Letters*, 20(8):813–815, 1995.

[DPNM+96] C. Deuerling, W. Prettl, M. Nuebler-Morritz, H. Niederdellmann, P. Hering, W. Falkenstein, and B. Bückle. *"Transmissions Systems for the Er:YAG-Laser (2.94 μm) - State of the Art" in Laser in der Medizin*. W. Waidelich (Ed.), Springer, Berlin, 1996.

[DPR86] U. Dürig, D. W. Pohl, and F. Rohner. Near-field optical-scanning microscopy. *Journal of Applied Physics*, 59(10):3313–3317, 1986.

[DRVM+97] B. O. Dabbousi, J. Rodriguez-Viejo, V. F. Mikulec, J. R. Heine, H. Mattoussi, R. Ober, K. F. Jensen, and M. G. Bawendi. (CdSe)ZnS core-shell quantum dots: Synthesis and characterization of a size series of highly luminescent nanocrystallites. *Journal of Physical Chemistry B*, 101(46):9463–9475, 1997.

[EB99a] S. Empedocles and M. Bawendi. Spectroscopy of single cdse nanocystallites. *Accounts of Chemical Research*, 32(5):389–396, 1999.

[EB99b] S. A. Empedocles and M.G. Bawendi. Influence of spectral diffusion on the line shapes of single CdSe nanocrystallite quantum dots. *Journal of Physical Chemistry B*, 103(11):1826–1830, 1999.

[ENSB99] S. A. Empedocles, R. Neuhauser, K. Shimizu, and M. G. Bawendi. Photoluminescence from single semiconductor nanostructures. *Advanced Materials*, 11(5):1243–1256, 1999.

[ER97] Al. L. Efros and M. Rosen. Random telegraph signal in the photoluminescence of a single quantum dot. *Physical Review Letters*, 78(6):1110–1113, 1997.

[FlGSL86] F. Favre, D. le Guen, J. C. Simon, and B. Landousies. External-cavity semiconductor laser with 15 nm continuous tuning range. *Electronics Letters*, 22(15):795–796, 1986.

[FLW99] X. Fan, S. Lacey, and H. Wang. Microcavities combining a semiconductor with a fused-silica microsphere. *Optics Letters*, 24(11):771–773, 1999.

[FPL+00] X. Fan, P. Palinginis, S. Lacey, H. Wang, and M. C. Lonergan. Coupling semiconductor nanocrystals to a fused-silica microsphere: a quantum dot microcavity with extremely high q factors. *Optics Letters*, 25(21):1600–1602, 2000.

[FSZ+00] L. Fleury, J.-M. Segura, G. Zumofen, B. Hecht, and U. P. Wild. Nonclassical photon statistics in single-molecule fluorescence at room temperature. *Physical Review Letters*, 84(6):1148–1151, 2000.

[GBS01] S. Götzinger, O. Benson, and V. Sandoghdar. Towards controlled coupling between a high-Q whispering-gallery mode and a single nanoparticle. *Applied Physics B*, 73:825–828, 2001.

[GBS02] S. Götzinger, O. Benson, and V. Sandoghdar. Influence of a sharp fiber tip on high-Q modes of a microsphere resonator. *Optics Letters*, 27(2):80–82, 2002.

[GBSU96] M. J. Gregor, P. G. Blome, J. Schofer, and R. G. Ulbrich. Probe-surface interaction in near-field optical microscopy: the nonlinear bending force mechanism. *Applied Physics Letters*, 68(3):307–309, 1996.

[GC97] J.-J. Greffet and R. Carminati. Image formation in near-field optics. *Progress in Surface Science*, 56(3):133–237, 1997.

[GD84] P. B. Gallion and G. Debarge. Quantum phase noise and field correlation in single frequency semiconductor laser systems. *IEEE Journal of Quantum Elecronics*, QE-20(4):343–349, 1984.

[GD96] C. Girard and A. Dereux. Near-field optics theories. *Reports on Progress in Physics*, 59:657–699, 1996.

[GDBS01] S. Götzinger, S. Demmerer, O. Benson, and V. Sandoghdar. Mapping and manipulating whispering-gallery modes of a microsphere resonator with a near-field probe. *Journal of Microscopy*, 202:117–121, 2001.

[GdSMB+04] S. Götzinger, L. de S. Menezes, O. Benson, D. V. Talapin, N. Gaponik, H. Weller, A. L. Rogach, and V. Sandoghdar. Confocal microscopy and spectroscopy of nanocrystals on a high-Q microsphere resonator. *accepted for publication in Journal of Optics B*, 2004.

[GdSMM+03] S. Götzinger, L. de S. Menezes, A. Mazzei, O. Benson, D. V. Talapin, N. Gaponik, H. Weller, A. L. Rogach, and V. Sandoghdar. Controlled coupling of a single emitter to a single mode of a microsphere: where do we stand? *Proceedings of SPIE, Laser Resonators and Beam Control VI*, 4969:207–214, 2003.

[GI94] M. L. Gorodetsky and V. S. Ilchenko. High-Q optical whispering-gallery microresonators: precession approach for spherical mode analysis and emission patterns with prism couplers. *Optics Communications*, 113:133–143, 1994.

[GI99] M. L. Gorodetsky and V. S. Ilchenko. Optical microsphere resonators: optimal coupling to high-Q whispering-gallery modes. *Journal of the Optical Society of America B*, 16(1):147–154, 1999.

[GKH+01] G. R. Guthöhrlein, M. Keller, K. Hayasaka, W. Lange, and H. Walther. A single ion as a nanoscopic probe of an optical field. *Nature*, 414:49–51, 2001.

[GKL61] C. G. B. Garret, W. Kaiser, and W. L. Long. Stimulated emission into optical whispering modes of spheres. *Physical Review*, 124:1807–1809, 1961.

[Göt98] S. Götzinger. *Gewebeablation durch Erbium-Laserstrahlung geführt durch flexible Transmissionssysteme.* Wissenschaftliche Prüfungsarbeit für das Lehramt an Gymnasien, Universität Kaiserslautern, 1998.

[GPI00] M. L. Gorodetsky, A. D. Pryamikov, and V. S. Ilchenko. Rayleigh scattering in high-Q microspheres. *Journal of the Optical Society of America B*, 17:1051–1057, 2000.

[GR00] I. S. Gradshteyn and I.M. Ryzhik. *Table of Integrals, Series, and Products.* Academic Press, San Diego, sixth edition, 2000.

[GRGH83] P. Goy, J. M. Raimond, M. Gross, and S. Haroche. Observation of cavity-enhanced single-atom sponaneous emission. *Physical Review Letters*, 50(24):1903–1906, 1983.

[GRL+03] S. Gulde, M. Riebe, G. P. T. Lancaster, C. Becher, J. Eschner, H. Häffner, F. Schmidt-Kaler, I. L. Chuang, and R. Blatt. Implementation of the Deutsch-Jozsa algorithm on an ion-trap quantum computer. *Nature*, 421:48–50, 2003.

[GSI96] M. L. Gorodetsky, A. A. Savchenkov, and V. S. Ilchenko. Ultimate Q of optical microsphere resonators. *Optics Letters*, 21(7):453–455, 1996.

[GTR+02] N. Gaponik, D. V. Talapin, A. L. Rogach, K. Hoppe, E. V. Shevchenko, A. Kornowski, A. Eychmüller, and H. Weller. Thiol-capping of CdTe nanocrystals: An alternative route to organometallic synthetic routes. *Journal of Physical Chemistry B*, 106(29):7177–7185, 2002.

[Hec01] E. Hecht. *Optik*. Oldenbourg, München, 2001.

[IAB+99] A. Imamoğlu, D. D. Awschalom, G. Burkard, D. P. DiVincenzo, M. Sherwin, and A. Small. Quantum information processing using quantum dot spins and cavity QED. *Physical Review Letters*, 83(20):4204–4207, 1999.

[IYM99] V. S. Ilchenko, X. S. Yao, and L. Maleki. Pigtailing the high-Q microsphere cavity: a simple fiber coupler for optical whispering-gallery modes. *Optics Letters*, 24(11):723–725, 1999.

[Jac98] J. D. Jackson. *Classical electrodynamics*. John Wiley & Sons, New York, 3rd edition, 1998.

[Kal02] T. Kalkbrenner. *Charakterisierung und Manipulation der Plasmon-Resonanz eines einzelnen Gold-Nanoparticles*. Dissertation, Universität Konstanz, 2002.

[KCJB97] J. C. Knight, G. Cheung, F. Jacques, and T. A. Birks. Phase-matched excition of whispering-gallery-mode resonances by a fiber taper. *Optics Letters*, 22(15):1129–1131, 1997.

[KDS+95] J. C. Knight, N. Dubreuil, V. Sandoghdar, J. Hare, V. Lefèvre-Seguin, J. M. Raimond, and S. Haroche. Mapping whispering-gallery modes in microspheres with a near-field probe. *Optics Letters*, 20(14):1515–1517, 1995.

[KDS+96] J. C. Knight, N. Dubreuil, V. Sandoghdar, J. Hare, V. Lefèvre-Seguin, J. M. Raimond, and S. Haroche. Characterizing whispering-gallery modes in microspheres by direct observation of the optical standing-wave patern in the near-field. *Optics Letters*, 21(10):698–700, 1996.

[KHS+01] S. Kühn, C. Hettich, C. Schmitt, J.-Ph. Poizat, and V. Sandoghdar. Diamond color centres as a nanoscopic light source for scanning near-field optical microscopy. *Journal of Microscopy*, 202:2–6, 2001.

[KK97] B. Knoll and F. Keilmann. Scanning microscopy by mid-infrared near-field scattering. *Applied Physics A*, 66(5):477–481, 1997.

[Kle81] D. Kleppner. Inhibited spontaneous emission. *Physical Review Letters*, 47(4):233–236, 1981.

[Kra02] P. Kramper. *Mikroskopie und Spektroskopie an photonischen Kristallen: Einschluß von Licht auf Subwellenlängen-Bereiche.* Dissertation, Universität Konstanz, 2002.

[KRMS01] T. Kalkbrenner, M. Ramstein, J. Mlynek, and V. Sandoghdar. A single gold particle as a probe for apertureless scanning near-field optical microscopy. *Journal of Microscopy*, 202:72–76, 2001.

[KS99] F. K. Kneubühl and M. W. Sigrist. *Laser*, volume 5. Teubner, Stuttgart, Leipzig, 1999.

[KSV02] T. J. Kippenberg, S. M. Splillane, and K. J. Vahala. Modal coupling in traveling-wave resonators. *Optics Letters*, 27(19):1669–1671, 2002.

[LBG+00] B. Lounis, H. A. Bechtel, D. Gerion, P. Alivisatos, and W. E. Moerner. Photon antibunching in single CdSe/ZnS quantum dot fluorescence. *Chemical Physics Letters*, 329:399–404, 2000.

[LC94] H.-B. Lin and A. J. Campillo. cw nonlinear optics in droplet microcavities displaying enhanced gain. *Physical Review Letters*, 73(18):2440–2443, 1994.

[LIHM84] A. Lewis, M. Isaacson, A. Harootunian, and A. Muray. Development of a 500 A spatial resolution light microscope. *Ultramicroscopy*, 13(3):227–231, 1984.

[LL93] K. Lieberman and A. Lewis. Simultaneous scanning tunneling and optical near-field imaging with a micropipette. *Applied Physics Letters*, 62(12):1335–1337, 1993.

[LLL+00] B. E. Little, J.-P. Laine, D. R. Lim, H. A. Haus, L. C. Kimerling, and S. T. Chu. Pedestal antiresonant reflecting waveguides for robust coupling to microsphere resonators and for microphotonic circuits. *Optics Letters*, 25(1):73–75, 2000.

[LLY92] C. C. Lam, P. T. Leung, and K. Young. Explicit asymptotic formulas for the positions, widths, and strengths of resonances in Mie scattering. *Journal of the Optical Society of America B*, 9(9):1585–1592, 1992.

[Lou97] R. Loudon. *The Quantum Theory of Light.* Oxford University Press, Oxford, 2 edition, 1997.

[MBB+03] J. McKeever, A. Boca, A. D. Boozer, J. R. Buck, and H. J. Kimble. Experimental realization of a one-atom laser in the regime of strong coupling. *Nature*, 425:268–271, 2003.

[MHG+01] G. Messin, J. B. Hermier, E. Giacobino, P. Desbiolles, and M. Dahan. Bunching and antibunching in the fluorescence of semiconductor nanocrystals. *Optics Letters*, 26(23):1891–1893, 2001.

[MHMS00] J. Michaelis, C. Hettich, J. Mlynek, and V. Sandoghdar. Optical microscopy using a single-molecule light source. *Nature*, 405:325–327, 2000.

[Mie08] G. Mie. Beiträge zur Optik trüber Medien, speziell kolloidaler Metallösungen. *Annalen der Physik*, 25(4):377–445, 1908.

[MIM+00] P. Michler, A. Imamoğlu, M. D. Mason, P. J. Carson, G. F. Strouse, and S. K. Buratto. Quantum correlation among photons from a single quantum dot at room temperature. *Nature*, 406:968–970, 2000.

[MK89] W. E. Moerner and L. Kador. Optical detection and spectroscopy of single molecules in a solid. *Physical Review Letters*, 62(21):2535–2538, 1989.

[MKB+00] P. Michler, A. Kiraz, C. Becher, W. V. Schoenfeld, P. M. Petroff, L. Zhang, E. Hu, and A. Imamoğlu. A quantum dot single-photon turnstile device. *Science*, 290:2282–2285, 2000.

[MNB93] C. B. Murray, D. J. Norris, and M. G. Bawendi. Synthesis and characteristion of nearly monodisperse CdE (E=S, Se, Te) semiconductor nanocrystallites. *Journal of the American Chemical Society*, 115(19):8706–8715, 1993.

[Moe02] W. E. Moerner. A dozen years of single-molecule spectroscopy in physics, chemistry, and biophysics. *Journal of Physical Chemistry B*, 106(5):910–927, 2002.

[NDB+96] M. Nirmal, B. O. Dabbousi, M. G. Bawendi, J.J. Macklin, J. K. Trautman, T. D. Harris, and L. E. Brus. Fluorescence intermittency in single cadmium selenide nanocrystals. *Nature*, 383(802-804), 1996.

[NERB96] D. J. Norris, Al. L. Efros, M. Rosen, and M. G. Bawendi. Size dependence of exciton fine structure in cdse quantum dots. *Physical Review B*, 53(24):16347–16354, 1996.

[NSW+00] R.G. Neuhauser, K. T. Shimizu, W. K. Woo, S. A. Empedocles, and M. G. Bawendi. Correlation between fluorescence intermittency and spectral diffusion in single semiconductor quantum dots. *Physical Review Letters*, 85(15):3301–3304, 2000.

[Oht98] M. Ohtsu. *Near-Field Nano/Atom Optics and Technology*. Springer Verlag Tokyo,, 1998.

[OKN80] T. Okoshi, K. Kikuchi, and A. Nakayama. Novel method for high resolution measurement of laser output spectrum. *Electronics Letters*, 16(16):630–633, 1980.

[Paw95] J. B. Pawley. *Handbook of Biological Confocal Miroscopy*. Plenum Press, New York, 1995.

[PDL84] D. W. Pohl, W. Denk, and M. Lanz. Optical stethoscopy: Image recording with resolution $\lambda/20$. *Applied Physics Letters*, 44(7):651–653, 1984.

[PLS+99] O. Painter, R. K. Lee, A. Scherer, A. Yariv, J. D. O'Brien, P. D. Dapkus, and I. Kim. Two-dimensional photonic band-gap defect mode laser. *Science*, 284:1819–1821, 1999.

[Pur46] E. M. Purcell. Spontaneous emission probabilities at radio frequencies. *Physical Review*, 69:681, 1946.

[PW91] H. Patrick and C. E. Wieman. Frequency stabilization of a diode laser using simultaneous optical feedback from a diffraction grating and a narrowband Fabry-Perot cavity. *Review of Scientific Instruments*, 62(11):2593–2595, 1991.

[PY99] M. Pelton and Y. Yamamoto. Ultralow threshold laser using a single quantum dot and a microsphere cavity. *Physical Review A*, 59(3):2418–2421, 1999.

[Ray78] J. W. S. Rayleigh. *Theory of sound*, volume 2. Macmillan and Co., 1878.

[Ray10] J. W. S. Rayleigh. The problem of the whispering gallery. *Philosophical Magazine*, XX:1001–1004, 1910.

[Rol99] G. Roll. *Optische Mikroresonatoren: Beschreibung im Bild der geometrischen Optik*. Dissertation, Ruhr-Universität Bochum, 1999.

[RTB+91] G. Rempe, R. J. Thompson, R. J. Brecha, W. D. Lee, and H. J. Kimble. Optical bistability and photon statistics in Cavity Quantum Electrodynamics. *Physical Review Letters*, 67(13):1727–1730, 1991.

[RTS+02] A. L. Rogach, D. V. Talapin, E. V. Shevchenko, A. Kornowski, M. Haase, and H. Weller. Organization of matter on different size scales: Monodisperse nanocrystals and their superstructures. *Advanced Functional Materials*, 12(10):653–664, 2002.

[San01] V. Sandoghdar. *"Trends and Developments in Scanning Near-field Optical Microscopy" in Proceedings of the International School of Physics ≪Enrico Fermi≫ course CXLIV*. M. Allegrini and N. García and O. Marti (Eds.), IOS Press, Amsterdam, 2001.

[SB91] S. Schiller and R. L. Byer. High-resolution spectroscopy of whispering-gallery modes in large dielectric spheres. *Optics Letters*, 16(15):1138–1140, 1991.

[SBK+03] V. Sandoghdar, B. Buchler, P. Kramper, S. Götzinger, O. Benson, and M. Kafesaki. *"Scanning Near-field Optical Studies of Photonic Devices" in Photonic Crystals - Advances in Design, Fabrication and Characterization.* K. Busch, S. Lölkes, R. Wehrspohn and H. Föll (Eds.), WILEY-VCH Berlin, 2003.

[SBN00] B. Sick, B.Hecht, and L. Novotny. Orientational imaging of single molecules by annular illumination. *Physical Review Letters*, 85(21):4482–4485, 2000.

[SBPM02] G. Schlegel, J. Bohnenberger, I. Potapova, and A. Mews. Fluorescence decay time of single semiconductor nanocrystals. *Physical Review Letters*, 88(13):(137401–1)–(137401–4), 2002.

[Sch93] S. Schiller. Asymptotic expansion of morphological resonance frequencies in Mie scattering. *Applied Optics*, 32(12):2181–2185, 1993.

[Sch96] R. Scheps. Upconversion laser processes. *Progress in Quantum Electronics*, 20(4):271–358, 1996.

[SHvdV98] M. Schader, S. W. Hell, and H. T. M. van der Voort. Three-dimensional super-resolution with a 4pi-confocal microscope using image restoration. *Journal of Applied Physics*, 84(8):4033–4042, 1998.

[Sie86] A. E. Siegman. *Lasers.* University Science Books, Sausalito, 1986.

[SM99] T. Saiki and K. Matsuda. Near-field optical fiber probe optimized for illumination-collection hybrid mode operation. *Applied Physics Letters*, 74(19):2773–2775, 1999.

[SPOS97] J. J. Scherer, J. B. Paul, A. OKeefe, and R. J. Saykally. Cavity ringdown laser absorption spectroscopy: History, development, and application to pulsed molecular beams. *Chemical Review*, 97(1):25–51, 1997.

[SPY01] G. S. Solomon, M. Pelton, and Y. Yamamoto. Single-mode spontaneous emission from a single quantum dot in a three-dimensional microcavity. *Physical Review Letters*, 86(17):3903–3906, 2001.

[STH+96] V. Sandoghdar, F. Treussart, J. Hare, V. Lefèvre-Seguin, J.-M. Raimond, and S. Haroche. Very low threshold whispering-gallery mode microsphere laser. *Physical Review A*, 54(3):R1777–R1780, 1996.

[Str41] J. A. Stratton. *Electromagnetic Theory.* McGraw-Hill Book Company, New York, 1941.

[Str99] E. Streed. *Spectroscopy of high-Qwhispering-gallery modes in silica microspheres.* Senior Thesis Experimental, California Intitute of Technology, 1999.

[Syn28] E. H. Synge. A suggested method for extending microsocopic resolution into the ultramicroscopic region. *Philosophical Magazine*, 6:356, 1928.

[TLL03] H. C. Tapalian, J.-P. Laine, and P. A. Lane. High-Q silica microsphere resonator sensors using stripline-pedestal anti-resonant reflecting optical waveguide couplers. *Proceedings of SPIE, Laser Resonators and Beam Control VI*, 4969:11–22, 2003.

[TS02] U. Tietze and Ch. Schenk. *Halbleiter-Schaltungstechnik.* Springer Verlag Berlin, 12 th edition, 2002.

[USH02] S. Uetake, R. S. D. Sihombing, and K. Hakuta. Stimulated Raman scattering of a high-Q liquid-hydrogen droplet in the ultraviolet region. *Optics Letters*, 27(6):421–423, 2002.

[Vah03] K. J. Vahala. Optical microcavities. *Nature*, 424:839–846, 2003.

[VFG+98] D. W. Vernooy, A. Furusawa, N. Ph. Georgiades, V. S. Ilchenko, and H. J. Kimble. Cavity QED with high-Q whispering-gallery modes. *Physical Review A*, 57(4):R2293–R2296, 1998.

[VIM+98] D. W. Vernooy, V. S. Ilchenko, H. Mabuchi, E. W. Street, and H. J. Kimble. High-Q measurements of fused-silica microspheres in the near infrared. *Optics Letters*, 23(4):247–249, 1998.

[Wal78] J. Walker. The amateur scientist-some whispering-galleries are simply sound reflectors, but others are more mysterious. *Scientific American*, 239:146–154, 1978.

[Wat22] G. N. Watson. *A treatise on the theory of Bessel functions.* Cambridge University Press, Cambridge, 1922.

[Web96] R. H. Webb. Confocal optical microscopy. *Reports on Progress in Physics*, 59:427–471, 1996.

[Weg98] S. Wegscheider. *Optische Strukturierung von Oberflächen ohne Beugungsbeschränkung.* Dissertation, Universität Konstanz, 1998.

[Wen94] Holger Wenz. *Aufbau eines Diodenlasers mit externem Resonator zur Amplituden-/Phasenmodulations-Spektroskopie.* Diplomarbeit, Universität Kaiserslautern, 1994.

[WH91] C. E. Wieman and L. Hollberg. Using diode lasers for atomic physics. *Rewiew of Scientific Instruments*, 62(1):1–20, 1991.

[WSH+95] D. S. Weiss, V. Sandoghdar, J. Hare, J.-M. Raimond V. Lefèvre-Seguin, and S. Haroche. Splitting of high-Q Mie modes induced by light backscattering in silica microspheres. *Optics Letters*, 20(18):1835–1837, 1995.

[WU02] A. Wallraff and A. V. Ustinov. Flüstende flussquanten. *Physik in unserer Zeit*, 33(4):184–190, 2002.

[WUK+00] A. Wallraff, A. V. Ustinov, V. V. Kurin, I. A. Shereshevky, and N. K. Vdovicheva. Whispering vortices. *Physical Review Letters*, 84(1):151–154, 2000.

[Wya85] R. Wyatt. Spectral linewidth of external cavity semiconductor lasers with strong, frequency-selective feedback. *Electronics Letters*, 21(15):658–659, 1985.

[ZEK02] Ch. Zander, J. Enderlein, and R. A. Keller. *Single Molecule Detection in Solution.* Wiley-VCH, Berlin, 2002.

[ZMW95] F. Zenhausern, Y. Martin, and H. K. Wickramasinghe. Scanning interferometric apertureless microscopy: Optical imaging at 10 Angstrøm resolution. *Science*, 269:1083–1086, 1995.

Danksagung

Nun ist es soweit. Ich kann mich endlich bei all denjenigen bedanken, die - auf welche Art auch immer - zum Gelingen dieser Arbeit beigetragen haben.
Herrn Prof. Dr. Oliver Benson möchte ich für sein Vertrauen danken, das er zu jedem Zeitpunkt in mich und in das Gelingen des Experiments gesetzt hat. Darüber hinaus nahm er sich immer Zeit - für persönliche Anliegen wie auch bei allen Problemen im Labor.
Natürlich gilt mein Dank auch Herrn Prof. Dr. Vahid Sandoghdar - er ist wahrlich ein Motivationskünstler! Selbst nach seinem Umzug nach Zürich blieb unser Kontakt sehr eng. So wurde er nie müde, mich von der Machbarkeit gewisser „einfacher“ Experimente zu überzeugen.
Nicht zuletzt geht mein Dank an Herrn Prof. Dr. Jürgen Mlynek. Er hat nicht nur dafür gesorgt, daß ich mich für das „richtige“ Experiment entschieden habe, sondern mir ebenso gezeigt, daß außerhalb des Labores vieles getan werden muß, damit Wissenschaft im Labor vernünftig funktioniert.
Die gesamte Konstanzer Gruppe hat mir ein sehr gutes Arbeitsklima beschert. Stellvertretend für die anderen Untergruppen seien Holger aus der Quantenmetrologie und Bernd aus der Atom-Optik genannt. Besonders Markus wünsche ich viel Erfolg mit seinem Team, wie er es gerne nannte, in Heidelberg. Er hat vor allem meine Freizeit in Konstanz recht angenehm gestaltet.
Mit Patrick durfte ich das Labor teilen. Wir hatten viel Spaß in der Dunkelheit des Labors. Wenn es darum ging Probleme mit der Verwaltung, den Behörden oder Versicherungen zu „lösen“ war er immer ein wichtiger und wertvoller Ratgeber. Thomas unterwies mich mit Geduld in der Kunst, entweder die Scherkraft oder mich selbst zu beherrschen. Dies war eine der wichtigsten Lektionen für mich als Anfänger auf dem Gebiet der Nano-Optik. Martin war zu jeder Tages- und vor allem Nachtzeit bereit, auch ausgefallene Dinge mit Lab-View zu programmieren. Die "Nanos" Patrick, Thomas, Christian, Carmen, Jan, Jens, Hannes, Sergei, Ilia, Philip und Lavinia waren nicht nur im Labor eine tolle Gruppe.
Der andere Thomas, einer der letzten echten Schwaben, mußte mich seit den ersten Tagen in Berlin fast täglich ertragen. Mit Stärke und Geduld hat er diese Prüfung bestanden. Die „Südländer“, Leonardo und Andrea, brachten Leben ins dunkle Microsphere-Labor. Beiden wünsche ich viel Erfolg bei der Weiterführung des Experiments! Aber auch allen anderen Mitglieder der Berliner „Nanos“ Kirstin, Felix, Val, Ounsi und Franz sorgten für eine Atmosphäre in der das Arbeiten viel Spaß machte.
In Zürich fühlte ich mich immer willkommen. Vor allem Sergei sorgte dafür, daß meine „Bestellungen“ immer pünktlich bei mir in Berlin ankamen.
Die Techniker und Ingenieure Stefan Eggert, Stefan Hahn und Klaus Palis haben durch ihre Fachkenntnis und Hilfsbereitschaft wesentlich zum Gelingen der Experi-

mente beigetragen.
Leonardo, Andrea und Glenn haben sich die Arbeit gemacht, diese Arbeit zu korrigieren.
Nicht vergessen möchte ich Dimitri und Nikolai von der Weller-Gruppe aus Hamburg. Beide versorgten mich vorzüglich mit Nanokristallen aller Art und Farbe.
Die Carl-Zeiss-Schott-Förderstiftung hat mich finanziell unterstützt - an dieser Stelle auch ein Dankeschön an diese Damen und Herren.
Alle meine Freunde haben stets dafür gesorgt, daß es eine schöne Zeit war. Insbesondere zu Hause in der Pfalz war man stets bemüht, mich etwas erleben zu lassen, wenn ich „auf Urlaub“ war.
Sehr wichtig für mich war meine Familie, auf die ich immer hundertprozentig zählen konnte. Ihnen gilt mein wärmstes „Dankeschön“.

Danke!

Selbständigkeitserklärung

Hiermit erkläre ich, die vorliegende Arbeit selbständig ohne fremde Hilfe verfasst und nur die angegebene Literatur und Hilfsmittel verwendet zu haben.

Ich habe mich anderwärts nicht um einen Doktorgrad beworben und besitze einen entsprechenden Doktorgrad nicht.

Ich erkläre die Kenntnisnahme der dem Verfahren zugrunde liegenden Promotionsordnung der Mathematisch-Naturwissenschaftlichen Fakultät I der Humboldt-Universität zu Berlin.

Berlin, den 11. November 2003

Stephan Götzinger